CROSS-FUNCTIONAL PRODUCTIVITY IMPROVEMENT

CROSS-FUNCTIONAL PRODUCTIVITY IMPROVEMENT

Ronald Blank

CRC Press
Taylor & Francis Group
Boca Raton London New York

CRC Press is an imprint of the
Taylor & Francis Group, an **informa** business

A PRODUCTIVITY PRESS BOOK

CRC Press
Taylor & Francis Group
6000 Broken Sound Parkway NW, Suite 300
Boca Raton, FL 33487-2742

© 2013 by Taylor & Francis Group, LLC
CRC Press is an imprint of Taylor & Francis Group, an Informa business

No claim to original U.S. Government works

Printed in the United States of America on acid-free paper
Version Date: 20120627

International Standard Book Number: 978-1-4665-1073-9 (Hardback)

Library of Congress Cataloging-in-Publication Data

Blank, Ronald.
    Cross-functional productivity improvement / Ronald Blank.
      p. cm.
    Includes bibliographical references and index.
    ISBN 978-1-4665-1073-9 (hbk. : alk. paper)
    1. Industrial productivity. 2. Production management. 3. Industrial management. I. Title.

HD56.B588 2013
658.5'03--dc23
                                                  2012025211

Visit the Taylor & Francis Web site at
http://www.taylorandfrancis.com

and the CRC Press Web site at
http://www.crcpress.com

For my professional colleagues everywhere, who by the nature

of their work strive daily to improve productivity

Contents

List of Figures

List of Tables

Preface

Productivity is strongly related to profitability, and making a profit is what business is all about. This is why businesses, whether manufacturing or service, need to operate productively. To survive in a competitive marketplace, high productivity must be maintained and even improved. But in the past 30 years there has been a change in the way companies operate. This necessitates a change in how they maintain and improve their productivity.

Traditionally, companies improved their productivity through automation and to a lesser extent by employee training. Then in the 1980s companies began to improve productivity by reducing their workforce, expecting fewer people to do more work. While this is a legitimate and effective way to improve productivity, it is actually counterproductive when overdone. Unfortunately, as the need for improved productivity increased, the companies became overreliant on this as a way to survive. In some companies, Lean Manufacturing has become corrupted and distorted from its original methodology and intent to become a philosophy of simply doing more work with less people.

Organizational management systems like ISO 9001 and its sector-specific variants like AS9100 and TS 16949 require companies to operate in a way that is integrated and interdisciplinary. To take full advantage of the benefits that these management systems have to offer, productivity must also be improved and maintained in an integrated, cross-functional way. These standardized systems require a high degree of integration and interaction between the various departments of a company. They also place value and emphasis on quality and organizational improvements. Thus quality has enjoyed a reprisal as an important productivity-enhancement tool. Six sigma and other quality methods are successfully used to improve quality.

However, due to the interdependencies, interactions, and high levels of integration fostered by the ISO 9001 system and it variants, maximum productivity is not achieved just by automation, training, leaner operation, and quality improvement alone. A more integrated, more all-encompassing approach involving all departments and a multifaceted approach to productivity improvement in manufacturing are necessary. Every aspect of company operations, including engineering, purchasing, production control planning, manufacturing, quality, shipping and receiving, and even human resources, among others, must do their fair share to improve the productivity of the company—and do so not as independent islands but in cooperation with all other departments. Only this cross-functional approach to productivity improvement can maximize the productivity of an organization. This book tells you how.

My intent is to explain how all the different elements of business operations can affect productivity and how each can do its part in improving the productivity of the organization. I reveal errors made by companies in various departments that can hurt productivity, but also provide practical solutions and alternatives to prevent or correct them. This book is not a cursory treatment. It provides the reader with the level of detail that would be expected in a textbook while explaining the knowledge and procedures necessary for the reader to implement cross-functional productivity actions of their own.

This book begins with an overview of productivity and traditional productivity improvement methods and then quickly expands to explain how the various departments can and do affect productivity. The subsequent chapters explain in more detail exactly how productivity is affected by the various departments and what must be done to have those departments improve productivity. Knowledge of the errors companies make that hurt productivity, and how to correct these counterproductive activities, is also included. Extensive chapters explain manufacturing methods and quality system improvements, which cover every aspect of manufacturing and quality as applied to maximizing productivity.

The figures in the chapters illustrate explanations, and the tables that are included provide the technical information necessary to implement productivity improvements.

The reader will finish this book with knowledge of what to do and what not to do in all aspects of company operations so that productivity can be maximized by implementing a truly cross-functional approach to productivity improvement. Furthermore, this book will enable the company to take better advantage of all that the ISO 9001 and similar systems have to offer by making good use of the interactions between the various elements of company operations.

At the end of the book is an extensive glossary that provides meanings of terms, some of which may be outside the reader's area of expertise. There is also a list of suggested readings for the reader to further explore productivity improvement.

About the Author

Ronald Blank, PhD, has worked in the automotive industry for 15 years and in aerospace for 12 years in addition to his years of experience as an industrial consultant for quality and productivity improvement. He holds a bachelor's degree and a doctor of engineering degree with a specialization in engineering management and quality control. Ronald Blank is the author of several books and technical papers on such topics as productivity improvement, reliability, internal quality auditing, and statistics. He has been a member of the American Society for Quality since 1980 and served on the executive board of the Hartford chapter. He lives in Middletown, Connecticut, where he works for an international engineering firm in the aerospace industry.

1

Basic Concepts

Productivity in industry is usually defined as the manufacturing output compared with an input. Output from the production process is compared with a chosen input and is usually expressed either as a ratio or as a percent. Labor productivity is typically measured as the number of production units of product manufactured per labor-hour. This is one example of a measurement of productivity. However, the input need not be labor-hours, as other aspects are relevant to productivity. Material productivity is the quantity of production output per quantity of material inputs. Total productivity can be calculated as total output quantity of units of product divided by total quantity of resources input. Any measure of productivity is simply the total saleable output compared to the chosen resource input, whether it is labor hours, material, or whatever.

Many publications discuss productivity, the measurement of productivity, distribution of productivity gains, and how to measure such gains. Whatever measurement of productivity is used, it must be one that will indicate increases or decreases in the productivity and might also include the distribution of the various production results among all the parties of interest.

When measuring productivity, only *saleable* production output should be considered. Rejected units of product are nonproductive. As the defect rate increases, the true productivity decreases, whereas an increase in real productivity represents a greater quantity of saleable product from the same amount of input—that is, the same amount of resources being consumed. Saleable production is also known as good product, conforming product, or shippable product. In an ideal situation, the productivity is determined at various stages throughout the entire manufacturing process.

This calls attention to the relationship between quality and productivity. Because defective products are not sold, the time and money spent producing them cannot be recovered. Except for recyclable materials, the materials consumed by defects cannot be recovered either. It is the loss of these nonrecoverable resources due to poor quality that reduces the productivity. Thus, improving quality, that is to say, lowering the proportion of defects, is a major factor in productivity improvement. It has about as much effect as changing the production rate.

The concept of total productivity requires measuring productivity as output divided by a variety of inputs, each giving a partial measure of total productivity. The number of units of product per labor-hour and the number of units of product per ton of raw material are two different measurements of productivity, yet each is a part of the overall productivity. These are called measurements of partial productivity.

Measurement of partial productivity refers to the measurement solutions that do not meet the requirements of total productivity measurement but are still practical as indicators of productivity. In practice, a measurement of production means a measurement of partial productivities. These measurements are components of total productivity. When interpreted correctly, these components are indicative of productivity development. The term *partial productivity* illustrates well the fact that total productivity is only measured partially (or approximately) and involves a variety of inputs.

Single-factor productivity refers to the measurement of productivity that is a ratio of output and one input factor. A very well-known measure of single-factor productivity is the measure of product output per labor-hour input.

Labor productivity is the ratio of production output to the input of labor. Where possible, the number of actual labor-hours worked, rather than the number of employees, is used as the measure of labor input. In situations where one employee may wear several hats in the company and where there is part-time employment, the number of labor-hours worked provides the more accurate measure of labor input. Labor productivity should be interpreted very carefully if used as a measure of efficiency. In particular, it reflects more than just the efficiency or productivity of workers. This is because labor output is influenced by factors that are outside of workers' influence, including the nature and amount of capital equipment that is available, the introduction of new technologies, and management practices.

When measuring productivity, a company may consider the productivity of a single work cell, an individual employee, a whole department, a product line, or even the entire company. It all depends on the purpose of the productivity measurement and how the productivity data are going to be used. When measuring the productivity of a whole company, all of the applicable production processes need to be analyzed to understand production performance.

However you measure productivity, your measurements must clearly convey the scope and base of the productivity measurement. If you measure productivity as total pieces per labor-hour, your measurement should explicitly say *labor-hour* so as to void any misconception or assumption that it might be pieces per unit of raw material or number of employees. If you measure the productivity of a given work cell or department, identify it to prevent any misunderstanding as to where the productivity was measured. This also ensures that the appropriate people get recognized for their productivity improvement. It may be necessary to have more than one productivity chart—one for each department or assembly line. Posting these together may cause some competition, which may increase employee motivation to improve productivity.

Productivity, no matter how it is measured—whether in relation to labor-hours, material, or any other chosen input—will naturally vary from day to day and from week to week. It most certainly changes from month to month and from year to year. That is why when you complete any productivity improvement action and you see what appears to be an improvement over a period of days or just a couple weeks, this is not the time to pat anyone on the back or rest on your laurels. There is enough normal variation on productivity to cause the appearance of short-term gain even if all your improvement actions were in fact totally ineffective. To really tell if your productivity improvement effort was successful, you would need to show a gain in the *average* productivity over a period of time. This gain must be beyond what can be explained by the normal random productivity variation. A moving average chart, plotted weekly with the productivity averages calculated on the previous four or five weeks, is a good way to check your productivity. Such a chart will absorb day-to-day variation and give you an average that will show increasing or decreasing trends when they occur. See Figure 1.1 for an example.

All production consumes resources. When your resource consumption rate has reached its maximum, further improvements in productivity cannot occur without a significant change taking place that enables a real change in resource availability and usage. This is as true of productivity measured in terms of labor as it is true of productivity calculated in terms of materials. Such a substantial change will typically involve a change in

Week	In pcs / hr	Moving Average	
1	518		
2	490		
3	525		
4	505	509.5	(average of weeks 1 through 4)
5	510	507.5	(average of weeks 2 through 5, etc.)
6	525	516.25	
7	505	511.25	
8	515	513.75	
9	521	516.5	
10	512	513.25	
11	530	519.5	
12	522	521.25	
13	508	518	
14	528	522	
15	530	522	

Plot of raw productivity data

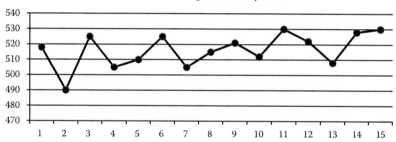

Moving average shows improvement trend more clearly

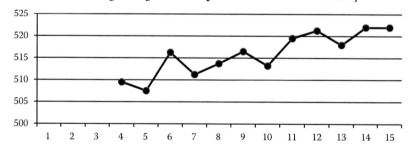

FIGURE 1.1

Moving-average chart for tracking productivity improvements.

material handling and trafficking, hiring practices, plant infrastructure, operating system, or a major process overhaul.

For any particular plant situation there is a maximum productivity. This is because labor output per hour, machine speeds, and all operational steps have a maximum output limit. How close you can come to achieving those maximums is variable—not only because of staffing variations for illness or other reasons of absenteeism, but also because of many other factors such as ambient temperature, employee experience, equipment aging and wear, and parts or materials availability, just to name a few. Besides the effect of your human resources, work cell layout, material handling practices, order of operations, and efficiencies of the procedures themselves all have an effect on productivity.

Knowing that there is a maximum productivity raises two questions: Can the productivity actually be improved? If so, by how much? This is determined by comparing the theoretical maximum production volume to the average actual production volume *of saleable product.* The more different these are, the more room you have to improve productivity and the easier it should be. As the actual average production volume of saleable product approaches the theoretical maximum, improving productivity will become increasingly more difficult, happen more gradually, and need to be more and more cross-functional.

However, the maximum productivity is not a constant value. New developments in technology and changes in employee demographics cause changes in the maximum productivity. Measuring productivity and comparing it to a 25-year-old benchmark is not a good idea. As things change through the decades, the maximum productivity can change also. Companies that do not change can fall behind in both productivity and market share. While everyone can think of exceptions where this is not the case, the truth is that companies that do not change not only stunt their own productivity but end up with markets that are more and more specific, which can stunt company growth.

Companies can increase productivity in a variety of ways. The most obvious and more traditional methods involve automation and computerization that minimize the tasks that must be performed by employees. Less obvious techniques are being employed that involve ergonomic design and worker comfort. A comfortable employee, the theory maintains, can produce more than a counterpart who suffers through the day.

Traditional methods of productivity improvement, though successful, tend to be of limited cross-functionality, if any at all. Typically these methods, although valid, involve a single approach to productivity and are of a more limited scope than they need to be. They do yield good results, but the results are unnecessarily limited. The productivity gains that are made by traditional and single-scope methods do not fully maximize productivity. Such limited scope methods include things like automation, employee productivity incentives, and material handling improvements.

Cross-functional productivity efforts not only are wider in scope, but they often result in greater productivity gains than traditional methods. Furthermore, they involve aspects and methods of productivity improvement that are not normally addressed in traditional improvement efforts. Cross-functional productivity methods more closely approach the productivity maximum.

The cross-functional approach to productivity includes almost all aspects of company activities—things like trainers and training methods, effects of purchasing practices, contract review technique, quality audits, design verification and validation, and even the effects of preventive maintenance. This multifaceted approach has been proven to achieve significantly higher levels of productivity. This does not mean to abandon the traditional productivity improvement methods. Rather it means to expand them by adding to them other aspects of company activities all in a way that is focused on improving the productivity of your product realization activities, that is, your production.

Do not misconstrue this cross-functional approach to mean that you simply make your overhead departments themselves more productive. It does not mean this at all. It means making your support departments more effective at contributing to productivity, and thereby enabling the company to achieve its maximum productivity. Cross-functional productivity improvement means that your support departments, normally considered overhead, can and do contribute to the productivity of manufacturing each in their own way. These effects can be subtle or somewhat obvious. They can be small or large, but they are often overlooked. They may even be ignored and unnoticed. Nevertheless, they are truly never insignificant. Indeed they are significant enough to make a real impact on the productivity of your manufacturing operation. Cross-functional productivity improvement also uses a wider variety of tools to improve productivity than traditional improvement techniques. Table 1.1 compares traditional productivity methods with cross-functional methods.

This book describes how various nonmanufacturing activities affect manufacturing productivity and how to have them contribute to productivity

TABLE 1.1

Comparison of Traditional and Cross-Functional Productivity Improvement

Traditional	Cross-Functional
Relies heavily on quality and manufacturing	Involves many different departments such as human resources, purchasing, and maintenance as well as manufacturing and quality
Emphasis on people and technique	Emphasis on tooling and equipment as well as people and technique
Little or no emphasis on preproduction activities	More emphasis on preproduction activities like contract review, purchasing, design verification and validation, and inspection planning
Minor use of preventive maintenance if any	Greater role of preventive maintenance as scheduled by actual data

improvement, but it does not ignore the more limited scope, traditional methods. In fact, to maximize productivity, you must take a combined approach. You must use every productivity tool and every method, and apply them well in order to maximize your productivity. The chapters that follow will tell you how.

Traditional approaches to productivity improvement work and produce real productivity gains, but they are limited in the amount of improvement that can be made. The reader can easily see that this book explains many different ways to improve productivity. This is because productivity is affected by so many aspects of business. With productivity improvement being affected by so many departments and company activities, how does one choose a particular improvement method? Maximum productivity improvement comes from an interdisciplinary, cross-functional approach. The chapters in this book tell you how to maximize productivity by expanding the scope of your productivity improvements to an interdisciplinary approach, so that it can be a fully cross-functional endeavor that will result in greater productivity gains from a variety of sources. Then the maximum productivity of the company can be achieved and the benefits of increased productivity fully realized.

2

The Traditional Approach to Productivity Improvement

To stay competitive in today's challenging industrial environment, companies have to be smarter in the way they do things. It is not enough to be lean and efficient. Companies need to operate wisely, that is, without making their own activities more difficult or less productive. Businesses can improve their productivity by planning carefully, breaking bad habits, taking new approaches to their activities, and not making some common errors.

Certain wisdom when applied to manufacturing companies can be of real benefit for increasing the productivity of manufacturing processes. Although such wisdom is to an extent a matter of knowing what to do and what not to do, it is also a matter of changing paradigms, breaking old habits, and implementing new ways of thinking.

Companies typically apply classical methods of improving productivity. Methods like six sigma quality improvement, Lean manufacturing, and others can and do result in real productivity gains. Nevertheless, companies seem to somehow still hinder their own productivity. It is often the culture of a company that stifles productivity improvement. Management style and technique, perceptions by the company leadership, and the company's operating system can all either help or hurt productivity. There are companies that truly want to be more productive, but because of the way they do things, they cannot seem to achieve the level of productivity they otherwise could.

Companies usually employ the classical methods of productivity improvement. Automation and computerization, which minimize tasks that must be performed by employees, are obvious methods for improving labor productivity; but less obvious techniques are also applied. These may involve things like ambient environment, safety equipment, hiring practices, even ergonomic design and worker comfort. A comfortable employee can produce more than an uncomfortable coworker who struggles through the day. Studies have demonstrated that work environment is of significant importance to productivity. Something as simple as raising or lowering the workplace temperature has an effect on productivity, even in an office environment.

No matter how much employees like the old ways, doing things differently can enable companies to stop interfering with their own productivity. By rethinking how a company does things, we can find and change what prevents productivity efforts from being as successful as they otherwise could be. In today's industrial environment there is no longer the degree of isolation between departments that there once was. Modern industrial operations have much more integrated systems, with a higher degree of interaction between departments. System management standards like ISO 9001 and its variants like AS9100 or TS 16949 describe fully integrated systems where various departments interact and affect the total productive outcome. This means that every department in some way affects the productivity of the company. Therefore, in addition to the classical productivity improvement methods, we may also identify more ways to improve productivity. This integrated, cross-functional approach will have major payback—and it will pay back sooner than you think.

First let's look at the traditional ways in which productivity is improved. These are all good and valid ways, but their effects will be specific and limited in scope.

EFFICIENCY AND LEAN MANUFACTURING

Improving efficiency is often seen as the way to improve productivity. Many people think that improving efficiency equals improving productivity. The truth is that improving efficiency often does result in productivity improvement, but it should not be relied on as the only way. This is because improving efficiency alone does not necessarily always result in as much improved productivity as could otherwise be achieved.

Measures of efficiency are actually very specific and are only one of several factors that may affect productivity. What this really means is that the specific efficiency of a specific process step, piece of equipment, and so forth is only one factor in the overall process capability. The efficiency of a particular step in a process or a particular piece of equipment can affect the overall productivity, but its effect is variable. In some cases it could have a significant impact on productivity. In other cases it may be minor or even negligible. A single improvement of efficiency may or may not be the most effective way to improve overall productivity.

An example might be the energy efficiency of a furnace used to melt alloys in a foundry. A more highly energy efficient furnace will melt the alloy sooner and with less energy and might indeed contribute to an improvement in productivity, but because other factors affect the productivity of the casting process, the increased furnace efficiency alone is not necessarily a *significant* improvement of the total productivity. Depending on the frequency of furnace use, fullness of the crucible, pour times, cast part cooling rates, and other processing factors, in some situations the effect on productivity will be variable. In some cases a more efficient furnace might have little effect on the productivity of certain products.

Lean manufacturing can either hurt or help. When properly implemented it can increase labor efficiency, lower labor cost, and improve productivity. However, when misunderstood or misapplied, Lean manufacturing actually lowers productivity.

Lean manufacturing is often misconstrued as simply doing more work with less people. Actually, Lean manufacturing says that to get the most productivity, you need to have the right people in the right numbers, doing the right things. Trimming away excess personnel so as to maximize the work output of the remaining personnel can make sense and improve productivity *when the right number of the right people remain and do the right things.* Merely reducing staff is not the proper way to do Lean manufacturing. As the number of employees decreases while still performing the same level of production, the efficiency of labor-hours goes up, but only to a certain extent. There is a point where the workforce output reaches a maximum for the number of people you have. Reducing the workforce beyond that point forces a reduction in productivity due to worker overload. Figure 2.1 shows how increased workload affects productivity.

Many books, training programs, seminars, and other resources are available to help you learn correct Lean manufacturing practices and increase productivity.

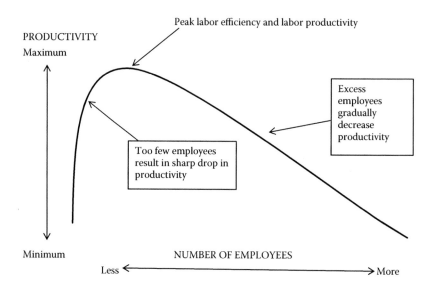

FIGURE 2.1
Productivity versus number of employees when workload is constant.

MATERIAL FLOW

Material flow can be thought of in two related ways: how materials flow through a plant, or the throughput of a specific operation. Productivity is improved by making either or both of these less wasteful. If you follow the pathway of material as it moves around in the factory floor, it is almost always zigzagging, crossing paths, and probably even has redundant moves. Less movement results in less time wasted moving and can also lower the cost of moving. Now think of how less movement can occur by rearranging where things are. Granted, completely changing the layout of the plant is a massive undertaking that may not be worth the effort for all factory sizes, but it is worth considering (and easier to do) in smaller companies. Any size plant can still benefit from this principle by rearranging the layout a specific department or even an individual work cell.

Increasing throughput of an individual operation can be a matter of rearranging the work cell. But it also pays to consider other things like container size, methods of material movement, and ergonomics. For an assembly operation where the operator has to take parts from bins and put them together, look for things like which bin empties first and how often it must be refilled. If an operator has to stop what they are doing to refill a bin, the bin size is too small. Each time they stop to refill the bin, it stops throughput on that operation. For maximum productivity in a day, one bin should hold the quantity needed for one shift's production. Then the operator need not ever stop production to go and get more parts.

When an operator gets parts from the previous operation to work on, how do they come to the operator? Does the operator have to go and get them? Are they brought to the workstation and the operator then reaches for them? Would it increase throughput if they came to the operator on a conveyer belt that the operator could control or by some other method?

The point is that the movement of material through the plant and the throughput of an operation are contributing factors to productivity. The decision not to change them is often a knee-jerk reaction justified by saying "It is too expensive to change" or "We can't shut the line down to make that change." These two excuses may be valid in some cases, but they ought to be examined with real time studies or real financial data. The decision should be based on actual cost and payback. Use actual cost when justifying the change to get a real picture of the effect. Don't rely on people's perceptions, opinions, and fears.

Different methods of moving material have different productiveness. Hand trucks or pushing wagons move material one load at a time. Conveyer belts move material continuously. Robotic vehicles move material all day with no breaks or lunchtime. They can also move trains or trolleys of many loads at once. For best productivity you must match the movement method to the way the material is used. Manufacturing operations that use materials at a constant rate and not in batches or lots should have constant movement of materials such as conveyer belts or other continuous feeding method. Operations that use material in lots or batches at a periodic time interval should have materials delivered in lots at matching intervals.

The just-in-time (JIT) material management method works well if the supplier quality and delivery are consistent enough. JIT systems bog down or even stop at irregularities if either item quality or delivery is not up to standard. Some companies mistakenly initiate JIT systems before their suppliers are ready or able to consistently provide the necessary quality or delivery. This mistake reduces productivity. They would be better off initiating the JIT system after they have worked with their suppliers enough to ensure consistent quality and delivery. Suppliers that are unable to provide acceptable quality or delivery should be given assistance to develop them to that point if practical; otherwise, consider changing supplers if JIT is to be implemented.

Kanban is another system for material movement. It is a *pull* system, meaning that the user pulls acceptable material as needed from a predetermined location, usually but not necessarily at or near the producer. The producer, noticing the missing material, then produces only the quantity of new material to fill up the Kanban storage area. This material is pulled through the manufacturing process rather than being manufactured ahead of time and pushed into storage areas in producer-determined lot sizes. Kanban moves material along at the *pace of usage, not the pace of production.* This prevents overstock. The amount of material on Kanban storage is kept constant by the producer, thus preventing shortages. Overstock and shortages both have a negative impact on true productivity. Kanban prevents this from happening and so improves productivity.

ERGONOMICS

The interaction between a person with their immediate environment, and equally important, the effects of those interactions on both the person and the product, can affect productivity. Ergonomic considerations also affect quality, which itself affects productivity. Consider such things as repetitive motion injuries, operator fatigue, left- and right-handedness, seating, operator height, age, strength, health, lighting, and so forth. All these things can affect productivity. Performing the operation yourself can give you realistic knowledge, but only if you do it long enough. Performing an operation for 5 or 10 minutes will not give you a realistic picture of how the operation and workstation affects the operator after a whole day or a whole week. Observe them at the start of their shift and then again about an hour before the end of their shift. Look for signs of sore muscles or hands, constant shifting of position, difficulty seeing, a decrease in the number of pieces per hour, and so forth as the day goes on. These are all indicators of ergonomics affecting productivity.

QUALITY

It costs the same amount of money and uses the same amount of time to build nonconforming product as it does to build good product. The difference is that the money spent building good product is returned to the company with a profit. This is obviously not true of nonconforming product, which has no payback or profit at all. Consequently, poor quality wastes resources, which cannot be recovered. It is not only a waste of time and material, but it is a waste of labor and a waste of the operating life of your equipment. Thus the lack of quality reduces productivity.

The importance of product quality is not an exaggeration. In fact, production quality is always a major influence on the productivity of any enterprise. However, a generalized quality improvement effort is not necessarily the best approach. Any quality professional can tell you that the 80–20 rule applies everywhere and always in the world of production quality. Therefore, to accomplish the most improvement, make a separate Pareto chart for each one of the following:

- Frequency of defects for each part number
- Frequency of defects for each defect type
- Frequency of defects by department
- Cost of defective part numbers
- Cost of defect types
- Cost of scrap and rework by department

Use these Pareto charts to steer the quality improvement effort to where it will have the most impact. Working where the Pareto charts steer your efforts will give you the greatest productivity improvement for the least effort and cost. Track these metrics monthly to see the success or failure of your quality improvement efforts.

There are many well-known quality improvement tools and methods, all of which, when successfully implemented, will enhance productivity. Six sigma projects, designed experiments, the define, measure, analyze, improve, control (DMAIC) methodology, and the like all contribute to productivity by improving quality.

However, because there are so many other aspects to productivity improvement, these must not be relied upon as the only productivity improvement activities that a company does. Emphasizing and utilizing only quality improvement as a means of productivity improvement could narrow a company's focus so much that all the other causes of low productivity might get overlooked and other productivity improvement methods ignored. Productivity improvements go beyond six sigma projects and other quality improvement methods.

Certain items are related to the quality system or practices that interrupt the normal flow of the process. Some of these can literally stop production. Others can adversely affect productivity by slowing down production. Looking more closely, we can see that quality issues affecting productivity are not just defective production but include things like poorly written procedures, conflicting requirements, and excessively specific system documentation. Other quality issues that affect productivity are improper use of management reviews, poorly done contract reviews, endless procedural loops, and poor document comprehensibility. Poor quality tools and substandard materials are also quality problems waiting to happen. Some may argue that these quality issues only affect quality. However, in most company operating systems, they can cause the flow of work to stall, or at least slow down, and that reduces productivity.

Identifying quality issues like these can be accomplished by thorough auditing, both internal and external; by doing document reviews; and by examining paper trails to find those that don't lead anywhere. Examples of some of the quality issues that require looking more closely to find are procedures that reference nonexistent documents, procedures that reference each other without giving the required details, forms or procedures requiring approval without identifying who has approval authority, and others. Forms have two purposes: they exist to tell someone what needs to be done, and they are a record of what was already done. Forms that do not give both of this needed information are themselves another quality and productivity issue.

Quality management system errors can affect any department, especially in a quality management system registered to ISO 9001 or one of its variants, because the quality procedures of such systems are highly integrated into everyday company activities, especially manufacturing. System planning errors and improper management reviews can prevent a company from fully benefiting from their registered quality systems. Audit traps written into the system are another common kind of error that adversely affects productivity by wasting time, labor, and money. See Chapter 5 for more information on how quality systems can affect productivity.

Quality affects productivity in other ways as well. Measurements and data collection are an important part of the quality function.

The measurements tell if the characteristics being measured are in tolerance or not, and that is the basis for distinguishing good product from nonconforming product. Data collection provides data that can be analyzed to foresee trends, identify differences, verify changes, and so forth. While these are good and necessary actions, they take time. Time spent inspecting is time not spent producing parts. So anything that can distinguish nonconforming parts from good parts and takes less time to do will improve productivity.

This is where attribute gauging comes in. Attribute gauges are go/no-go; the characteristic is either in tolerance or not. These kinds of measurements are faster and easier for operators to do. They are also often more error resistant. While it is true that they do not give the amount and kind of data for analysis that variables-type measurements can, they may still be worth doing. Attribute gauging can be set up for very rapid measurements when the right kind of gauges and fixtures are used. Custom attribute gauging can even check several characteristics at a time. When operators have to do inspection of parts they manufacture, this type of inspection can give operators more time to produce parts by spending less time inspecting.

MALPRACTICE

Malpractice actions are innocent or deliberate. Innocent malpractice is making mistakes without intended malevolence. Deliberate malpractice is the result of someone choosing to do what is wrong, with or without good intentions. Although actually rare, it is still more common than anyone wants to admit. Malpractice can be found not only by observing work activities and discussing them with employees, but also by reviewing and analyzing customer complaints, internal and external requests for corrective actions, discontinuities in records and workflow, and well-executed audits.

Innocent malpractice can be corrected by education and job training, which is sometimes viewed as a necessity due to one or more flaws in the system. Sometimes a highly motivated or quality-conscious employee simply does not know better and is trying to help the company or do their job very well. Innocent malpractice may be an operator tightening a nut with extra torque beyond what is specified so that it does not come off. Such operators are trying to help quality by making sure the nut does not come off, but in reality they are distorting the part, thereby ruining proper fit, or overstressing something, which may cause a premature failure.

Deliberate malpractice may simply be a cover-up or a personal favor to someone. It is not always done with malicious intent. Often the perpetrator has noble intentions. He or she almost always feels justified. An operator may "adjust" the numbers or "correct" the paperwork because they don't want to make waves, or to cover someone else's error, or because they want to make the part look good. This is deliberate malpractice.

Deliberate malpractice is also often viewed by the perpetrator as necessary due to one or more systemic inadequacies. In this case the person is compensating for a problem in a procedure or some other aspect of the company's operating system. They feel the compensation is necessary because in their perception the problem cannot be changed. Conflicting priorities expressed by management, conflicting quality requirements, and the perceived apathy of management are all rationalized by operators to justify deliberate malpractice without malicious intent. The need to perform the malpractice may be real or imaginary, and it may be true that the situation that motivates the malpractice cannot be changed. Whether the problem is real or not, the operator is deliberately making an unauthorized change, and by doing so, may reduce productivity. Even with good intentions this is deliberate malpractice. Unauthorized variations from what is supposed to happen can also be the result of someone in management misprioritizing or trying to compensate for misprioritization. Other intentional malpractice may be caused by malcontent employees or sabotage.

Regardless of the category or cause, these errors, malpractices, mistakes, or whatever you want to call them are responsible for the loss of millions of dollars, not to mention poor quality and the loss of customer goodwill. It behooves everyone in industry to identify and correct them, or better yet, to prevent them. Putting an end to any kind of malpractice will affect productivity for better or worse. When the underlying reason for the malpractice is truly corrected and not merely compensated for, then productivity will increase.

AUTOMATION

Automation is an often-used method of increasing productivity. Automating one or more process steps allows for more work to be accomplished by fewer people. It can enable employees to multitask more easily by freeing them from one process step so they can perform another. Automated processes can work faster and more consistently than people, and they do not tire, require time to eat or take breaks, and need any financial compensation or paid benefits.

Initially, this can seem like the perfect way to increase productivity. However, there are other things to consider that may make automation less attractive. This is not to say that automation is not a good idea. It certainly can be a good way to improve productivity, but all aspects and effects must be considered before making the decision to automate to a higher degree than you already have.

The obvious factor to consider is cost. The actual cost of building or purchasing the automated equipment plus the cost of running and maintaining it must be considered in comparison to the benefit of having and using the automation. But you must include all the costs. These may well include the following:

- Purchase and installation of equipment (if purchased)
- Design and assembly (if in-house)
- Electrical, pneumatic, and plumbing connections
- Operating costs
- Training costs
- Set-up and testing (time, materials, and personnel)
- Preventive maintenance
- Consumables (oil, coolant, filters, etc.)

Cost is certainly the most important and often deciding factor. Other factors to consider are available space, any necessary changes to the facility, time from conception to actual productive use, noise and other environmental impacts, frequency and ease of product changeovers, as well as the ability of the infrastructure and operating system to support the automation.

An important but often overlooked consideration is the reliability of the automated equipment. Reliability is the predicted probability that a device will operate as specified, under the specified conditions, for the specified period of time. How do the specified conditions for operating the automation compare to shop floor conditions? How does the specified period of time compare to the production requirements the automated equipment will be required to meet? The reliability must be good enough to economically meet production demands with little or no downtime due to failure.

Sharp drops within productivity caused by unexpected downtime might defeat any expected improvement in productivity.

The point being made here is to not assume that more automation is better. Rather, recognize that the right amount of the right kind of automation may improve productivity. However, many things need to be considered to determine the right amount and the right kind of automation.

Productivity improvement is an interdisciplinary subject. In addition to the traditional methods of productivity improvement, there are the effects of various departments and activities from aspects of company operations that have a supporting role to production, rather than being directly involved in it. These departments and activities also affect productivity. Actions performed by the purchasing department, decisions made by manufacturing and industrial engineers, improperly implemented statistical process control, human resource department activities, and poorly done contract reviews are just some of the different supporting activities that affect productivity. These topics are discussed in Chapters 4, 5, 6, and 8 in this book.

3

Additional Considerations for the Cross-Functional Approach

Productivity is actually a multidisciplinary goal. Many factors affect productivity and involve many departments. This book presents a comprehensive, multidisciplinary, cross-functional approach to productivity improvement. It reveals many different ways that all aspects of company operation and departmental activities can affect productivity. The secret to cross-functional productivity improvement is to not limit the improvement effort to one or even a few improvements made sequentially, but to do many of them. They may be done either sequentially or simultaneously as resources, corporate culture, and system limitations permit. They must be done in harmony with each other. A harmonious implementation is one where the implementation in one department does not hinder another's implementation, but rather they facilitate each other's changes. Activities included in this cross-functional approach include but are not limited to training, purchasing, sales contract review, quality auditing, and product measurements systems, among others.

While some of the sections of this chapter may seem to present obvious information about productivity, they also show how multifaceted productivity improvement can really be. This chapter is not all-inclusive, but it may start the reader into thinking outside the box and lead to a wider perspective on productivity improvement.

TRAINERS AND TRAINING METHODS

Many companies believe that the person best suited to train a new employee is someone who actually does the work the new employee is being trained to do or someone who has done it in the past. This is correct in that it passes down tribal knowledge and the hidden knacks and techniques not described in the instructions. It is a way to pass down subjective knowledge of what the new employee is supposed to do. It passes down the benefit of the previous employee's experience. However, it also passes down the bad habits and mistakes. It ensures the continuation of all the misunderstandings, bad habits, and misinformation accumulated by the employee doing the training.

Alternatively, some companies have a training department with trainers who may or may not have the benefit of practical experience and who probably never developed the hidden knack or technique. With a designated trainer you may lose the benefit of passing down the things that only experience can teach. If the designated trainer has not performed the job for any length of time, they may not have all the experiential knowledge. Their knowledge will be more theoretical, objective, and usually more in accordance with the written instructions. However, if your designated trainer is indeed well trained and experienced in the task being taught, then the benefit of a designated trainer is that you limit the passing down of the misinformation, bad habits, and misunderstandings, or at least limit additions to them.

Either choice, the on-the-job training by experienced employees or the training provided by a designated trainer, has its flaws and advantages. Consciously or unconsciously, deliberately or by default, companies effectively are making a choice of which flaws in training they want to live with whenever they decide how to train new employees. It is not unusual for the more experienced employee on the line to be the designated trainer, but when this is the case, management should be cautious about passing down bad habits and misunderstandings. A preemptive check with the person doing the training may be worth the time it takes.

Being trained "cookbook style" by following step-by-step instructions or by having an engineer or manager do the training has certain advantages, but is not necessarily the best way to train for every kind of activity. You lose the tribal knowledge, the hidden knacks and techniques not described in the instructions. You also don't pass down subjective knowledge of what the new employee is supposed to do and you lose the benefit of the previous employee's experience. You risk emphasizing the wrong points and teaching things that are not relevant to the day-to-day operation. You also may intimidate the employee being trained. Demands on management and engineering time could even result in feeding the new employee too much information at once or not transferring all the required information.

How then should an employee be trained so as to make the training most productive? How do you pass down the benefit of years of experience, practice, and tribal knowledge, while not passing down bad habits, misinformation, and misunderstandings, not to mention poor techniques that may hinder productivity? The answer is that you use a combination of training techniques, and if practical, a combination of trainers, all in the proper sequence, each doing their own particular part of the training. Care must be taken that the training from different trainers is connected, with no break in continuity of thought, and is smoothly overlapping.

For training on activities of moderate to high complexity or where a more formal training plan is called for, here is an example of how it's done: First present the theory; a manager or engineer would be good at this. Include the details of how and why, but do not actually demonstrate the operation; just merely explain it. Instruct the new employee this way in front of the previous operator and, if you have one, the designated trainer. The technical expertise of the manager or engineer can prevent misunderstandings and misinformation from being passed on.

Then the experienced employee or designated trainer demonstrates and teaches the operational technique. This is done in front of the engineer or manager, who is free to correct any error, bad practice, misinformation, or misunderstanding and comment on technique. Their presence can add importance and credibility to the training activity. Then the experienced operator begins the practical on-the-job training. On the first day when the newly trained employee does the job on their own, the engineer or manager and/or the designated trainer visit the new employee, and they always have the experienced employee with them during the visit. They observe the operation, but correct only if necessary. This team approach to training prevents the continuance of bad habits, misinformation, and poor technique.

Companies or individuals within a company may object to this. Most companies do not have designated trainers. Not every operation is so complex that such a level of team training is necessary. Three people training one person could cause confusion or even be intimidating. All these statements are true. But the principles of training for maximum productivity still apply. They can be applied with variation. Taking a closer, more realistic look at the principles and allowing some flexibility in method can result in efficient and effective training.

Most activities performed will not have the complexity to warrant such elaborate training. Simplified modifications of this plan will be more suitable in such cases as long as the basic principles are maintained. However you train employees, the benefit of previous experience and proper job knowledge must both be presented. Whenever practical, you must also pass on the hidden knack and subjective knowledge without passing on the bad habits and misunderstandings.

The training itself is that it must have a combination of theory and practical experience. One without the other will not make for effective and highly productive training. The three-person team mentioned above is just one plan for ensuring these principles are adequately addressed and fulfilled.

In many companies there is a lead person who has done every job in the department or at least most of them, is very competent, and is capable of doing any of the operations. That person can function as both the designated trainer and the experienced operator in the above-mentioned training team. Then only an engineer or manager needs to be added. The two-person team can provide good productive training. Alternatively, an engineer or manager can be the designated trainer. When coupled with the addition of an experienced operator, effective training can result.

No training is complete without an evaluation of how effective it is. This does not have to be a written test. Training is best evaluated by observing how well the newly trained person is doing the job. Ideally this evaluation should be done not only by the people who did the training, but also by person who receives the results of the new employees' work, that is, the subsequent operator. Who better to judge the quality of the just-completed operation than the person who must perform the next operation? They know best what the part or subassembly needs to have and what it is supposed to look like at that stage. The experienced operator and the subsequent operator are the best people to judge how effectively the person was trained. The effectiveness of the training is an important factor in how productive the individual is, as well as the productivity of the department as a whole. The evaluation of the training effectiveness must include input from the person or people involved in performing the training as well as examination of the expected result. Exactly how training and evaluation for training effectiveness is done in your company, and by whom, is less important.

EFFECTS OF PURCHASING ACTIVITIES

The purchasing function in a company has an impact on productivity. Obviously buyers need to consider costs, lead times, and on-time delivery records of the suppliers; but these are not the only factors affecting productivity. Poor quality causes rejects at incoming inspection. If not caught there, poor quality can result in defective product during manufacturing, or even final inspection. Whenever any nonconformance on a purchased item occurs, it reduces productivity. It wastes time while items are put on hold awaiting disposition. It wastes money in the forms of labor costs and material scrap costs. It decreases productivity by requiring rework, replacement, or repair of manufactured goods.

Purchasing, by placing too much emphasis on costs and delivery and not enough on quality, may be a contributor to the reduction in productivity caused by poor quality. The key is to have a proper balance of costs, delivery, and quality. When the amount of money applied to repair or rework purchased items is greater than the amount of money saved by purchasing from a low-cost supplier or a short lead time supplier, then the balance between costs, delivery, and quality is not as it should be. The money saved is then spent by having to deal with the quality issues.

Buyers will not necessarily know how their choice of supplier is affecting productivity. Here is where good communication between the purchasing department and manufacturing and quality departments is needed. A buyer may think they are saving money by ordering from a specific supplier, but due to lower quality, they may actually be costing the company more money than they are saving due to time and resources spent on repairs and rework. They would not know this unless the information is brought to their attention. They need to be well informed to create and maintain the proper balance between costs, delivery, and quality.

Another way that purchasing affects productivity is by ordering from unapproved suppliers, especially if they do not tell the quality department. Some companies have approved suppliers from which commodities and services must be purchased. These suppliers are often approved by either the quality department or the purchasing department, or both, according to the requirements of the management system of the company. But what happens when an order is placed from an unapproved supplier and the parts are delivered before the supplier gets approved? In most companies this would require either a deviation or a very much increased inspection, or even a 100 percent product sort. Any one of these is expensive and time consuming. The time and money used on these activities decreases productivity, and the added inspection cost may outweigh any cost savings from the unapproved supplier.

Suppose the order was placed with an unapproved supplier, and before the parts are delivered a supplier audit disqualifies the supplier. What then? If the order is allowed to ship, heavy inspection or sorting is again needed, thereby wasting time and money while decreasing productivity.

It is recognized that sometimes a buyer has to order from an unapproved supplier to meet production schedules or to keep costs within budget. Occasional business exceptions are inevitable in today's markets, but when these situations are too frequent, it is an indication of more serious problems in the way the company operates and ought to be addressed by the management team.

CONTRACT REVIEW TECHNIQUE

Contract review is a requirement of the ISO 9001 standard and its variants such as those standards for the automotive and aerospace sectors. It is how customer requirements are identified, planned for, and get translated into work orders and instructions. How this is done can have significant impact on productivity. Companies do contract review in various ways, ranging from one person, usually a salesperson, verifying things like items ordered, quantities, prices, and delivery dates to performing a comprehensive review by a cross-functional team composed of management-level employees from various departments including quality, manufacturing, engineering, and production control.

Having contract review being done by one person can have an undesirable effect on productivity. This is because one person does not have all the practical expertise to do a review as comprehensively and effectively as a cross-functional team can. Except in extremely small companies, a contract review that is performed by just one person hinders planning ahead by the departments that have to do the actual work to realize the product. One person is usually not as good at problem identification and situational preparedness within the various departments as a cross-functional team would be. Consequently problems that were not identified, as well as the lack of preparedness, must be dealt with during production. Being less prepared and having to solve or compensate for problems during production reduces productivity because it forces the diversion of resources and may use up labor time. The inability to do proper planning ahead on preventive maintenance, calibration, training, and so forth can also cause delays and increase production time, which always results in lower productivity.

The advantage of this cross-functional approach to contract review is that it enables planning ahead to maximize productivity. Activities like preventive maintenance and instrument calibrations can be rescheduled so as to minimize their impact on production. Training needs can be predetermined and training completed on time, before it is needed. The need for new tooling or fixturing becomes known sooner, so tools arrive and are available when production begins. Figure 3.1 is an example of a cross-functional contract review checklist, which also serves as a record that the contract review was done.

```
SALES:                    Sales Signature_____
Customer Name_____  Order Number_____  Date_____
Part Number Ordered            Quantity            Quoted Ship Date

_____   _____   _____
_____   _____   _____
_____   _____   _____


ENGINEERING:          Engineering Signature_____
New part?                       Y      N      Specify_____
New/revised drawings needed?    Y      N      Specify_____
Additional specifications needed?   Y   N     Specify_____


MANUFACTURING:        Manufacturing Signature_____
New work cell/assembly line needed?   Y   N   Specify_____
Secondary operations needed?    Y      N      Specify_____
Additional tooling/equipment needed?  Y   N   Specify_____
Additional personnel needed?    Y      N      Specify_____
Additional training needed?     Y      N      Specify_____


QUALITY:              Quality Signature_____
Inspection per _____   Sampling plan_____
Special testing per_____
Additional measuring equipment needed?  Y   N   Specify_____
Additional staffing needed?     Y      N      Specify_____
Additional training needed?     Y      N      Specify_____
```

FIGURE 3.1
Example of cross-functional contract review form.

INTERNAL AUDITS

Internal audits that serve merely an enforcement role can either help or hinder productivity. Enforcing the prescribed way of doing something can help productivity only when the prescribed way is actually more productive than what would otherwise be done without the enforcement. However, if the prescribed way is less productive, the audit may actually interfere with productivity by enforcing a less productive process method. Even if the prescribed way is a better way when considered from a certain perspective, it is not necessarily the most productive. When viewed in a larger, more long-term context, it may or may not actually be a better way. What should happen if a more productive way is being done but is not the prescribed way? In this situation you must always consider the fact that if operators are repeatedly doing something in a way that is different from the prescribed way, there must be a reason, valid or not.

Before assuming the prescribed procedure is always best, examine why the operators resist doing it. Is it unnecessarily prescriptive by trying to describe every aspect of how and when each activity is done in too much detail? Is it something that has not been reviewed and revised in so long a time that it is now no longer current? Does it make the best use of existing tooling and equipment? Is it worded in engineering language and too technical for operators to understand? Is it one person's personal method and not necessarily best for everyone? Answers to questions like these can shed some light on why a prescribed method is being unofficially replaced by a different method. Examine the merits of the operator's way and discuss the pros and cons of each method with the operators and with manufacturing or industrial engineers (or whoever fills that role in your company). The operators are the ones who do the procedure every day, all day long, so they may know best how to do it productively. To arrive at the best method, you may need to give them input on knowledge of engineering, quality, and reliability in your discussion. All of these factors need to be considered in deciding which way is best.

In the above scenario, the internal audit would not be used merely as a policing or enforcement action. It would be used as a starting point to reevaluate how an operation should be done to maximize its productivity.

Audits can be helpful or not in other ways, too. A good audit answers not only the typical questions "Do you have a procedure" and "Are you following it," but also "Is the procedure meeting the company's needs" and

"Is the goal of the procedure being accomplished?" Such an audit can be the trigger that initiates the fine tuning of a less productive procedure into a more productive one.

MEASURING SYSTEMS

Measurement systems must be up to the task to which they are applied. In fact, ensuring that the measurement system for any particular characteristic is truly adequate is an ISO 9001 requirement. Inadequate measuring systems affect productivity because they are a source of variation that can cause either of two problems. One problem is that they can reject material that should have been accepted and therefore artificially increase the defect rate. This forces more processing to be done to yield the same number of good parts, which obviously hurts productivity. The second problem is that they can accept something that should have been rejected at the current operation, only to be rejected further in the process when scrapping something is more expensive or, even worse, shipped to the customer. Either way, the error caused by measurement variation lowers productivity. The amount of variation induced by the measurements themselves is often considered to be very small, but in reality it often is 20 to 50 percent or more of the total tolerance window. Measurement system analysis tells you exactly how the measurement system increases variation in the characteristics being measured. It tells you the sources of variation and how much of the total variation is due to measurement alone, and gives information on what to do about it.

First, let's define what is meant by a measurement *system* as opposed to a measuring *device*. The measuring device is simply the tool you use to do the measuring, whereas the measuring system is all the components involved in making the measurement including the device, the person doing the measuring, the part being measured, and the environment in which the measurement is being made.

The measuring device itself has bias, which is the degree in which the measured value deviates from the actual true value. This is a matter of calibration and can be so little as to be imperceptible and insignificant, or so large as to require the device to be adjusted, repaired, or replaced. The device will also have linearity. This is the change in bias over the full range of the measurements that the device is capable of measuring. Linearity is the reason why devices must be calibrated over their full range. Linearity can be easily determined by plotting the deviation from known value (bias) over a series of checks along the whole range of the device. The degree of straightness you have on this plot is the linearity. Any variation in bias direction or any anomalous point indicates a nonlinear condition.

Stability of the calibration of the device is how the calibration changes over time. This, too, can be essentially imperceptible and insignificant, or great enough to cause a problem. It is normally plotted on a graph which is then treated like a statistical process control (SPC) chart. Stability is determined from the graph in the same manner that an SPC chart tells you the stability of a process (see Chapter 6 in the section on SPC). Figure 3.2 shows a linearity determination and Figure 3.3 shows a stability determination.

The combination of the device and the person doing the measuring has repeatability. This is the degree of variation observed when the same person is using the same device in the same way, making the same measurements on the same part multiple times. The width or range of the distribution of data obtained this way is the repeatability of the combination of person and device.

These characteristics of a measuring device are intrinsic to the kind of device being used. Indeed they are intrinsic to a particular device, not just a device type. But they are also influenced by the environment. Things like vibration, cleanliness, improper storage, age, and temperature, and so forth can all influence the bias, linearity, stability, and repeatability of the measuring device. The measurement technique of the person and the condition of the measuring device also determine the repeatability of the measurement. The more the measurements are influenced, the more variation they will induce in the measurement.

Operators themselves also are a source of measurement variation. Differences between operators using the same measuring device will have different values and levels of repeatability, depending on their technique and skill in using the device. In an ideal situation, it should not matter which operator makes a measurement. They all ought to have the same results. That is to say, each operator should be able to reproduce the results of the other operators. The degree to which the results of each operator differ on average is called the measurement *reproducibility*. It is a measure of the ability of all the operators to get the same results. This can be measured by having each operator measure the same sample of parts two or three times and calculating their averages. The maximum difference between one operator's average and another's is the reproducibility.

Gauge is a 6" caliper so the measurement range is 6"
Measure the bias over the whole range of the gauge

Print measurement is 5.665" ± .002"   Total tolerance is .004"
Desired linearity is < 30%

Measurement	Bias
.100"	0.0001
.500"	0.0001
1.000"	0.0002
3.000"	0.0001
5.000"	0.0001
5.500"	0
5.900"	0.0001
6.000"	0.0001

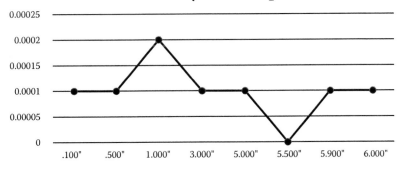

Linearity of 6" caliper

Linearity is the change in bias over the entire range, here shown as 0.0002"
which is 5% of the total tolerance so linearity is acceptable.

FIGURE 3.2
Gauge linearity example.

The resulting total (vector sum) of the repeatability and reproducibility of a device is known as gauge R&R, or GRR. Calculating GRR is the standard way of measuring the amount of variation in the measurements that is due to the measuring process itself. To put it another way, it is a method by which to determine the degree or probability that the measurement system itself is giving false readings and therefore is accepting bad parts or rejecting good ones. The GRR is often expressed as a percent of the total tolerance window. This total tolerance window is the upper specification limit minus the lower specification limit. The GRR is a measure of the suitability of the measuring system for the intended measurement.

Measurement device is a pressure gauge   Measurement interval is 1 week

Total tolerance: .0004"
Desired stability is < 30% of the total tolerance

Week	Bias
1	0
2	0.0001
3	0.0001
4	0
5	-0.0001
6	0
7	-0.0001
8	0
9	0.0001
10	-0.0001
11	0
12	0.0001
13	-0.0001
14	-0.0001
15	0

Stability over 15 weeks

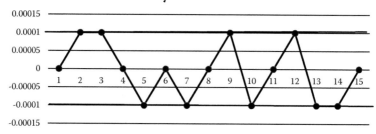

Pressure gauge is shown here to be unstable over 15 weeks because the bias variation is ± .0001 which is 50% of the tolerance. Since the desired stability is < 30%, the stability is not acceptable.

FIGURE 3.3
Gauge stability example.

A high GRR percentage means the measuring system is less suitable for use. Several publications describe how to measure GRR.

How much measurement-induced variation does it take to influence productivity? As a rule of thumb, a GRR that is greater than 30 percent of the total tolerance window is problematic. Some industries use 20 percent of the tolerance window as a maximum allowable GRR. To ensure that the GRR is not hurting your productivity, it should be kept at about 10 percent or less. Many companies maintain a policy that GRR less than

10 percent of the total tolerance window for the characteristic being measured is always acceptable. Between 10 and 20 percent may be acceptable due to limits of technology or cost, and up to 30 percent is acceptable only with approval from a competent authority, who is often either the customer or the company quality engineer or manager. Greater than 30 percent is not acceptable and requires either a corrective action or a change in the measurement system. Often a different device and technique are needed to improve GRR. It is important to improve GRR when it is too great because a high GRR can result in rejecting good product, which unnecessarily reduces productivity. Accepting nonconforming product also affects productivity because a nonconforming characteristic that is inadvertently accepted may cause a defect or test failure later on the in the process. It will be a more expensive defect then because more material, labor, and time have been invested by the time the defect is found. Table 3.1 tells how to remedy the various sources of variation when GRR is not good enough.

Process capability determines how sensitive a characteristic is to measurement-system-induced variation. The higher the capability, the less sensitive the characteristic will be to measurement variation. When attempting to improve a low capability, a good way to begin is by checking the GRR. If it is more than 20 or 30 percent of the tolerance window, you should definitely try to improve it. A lower GRR is always better. Improving (lowering) GRR to cause less measuring-induced variation will make a real improvement to process capability and that inevitably results in improved productivity.

TABLE 3.1

Remedies for Poor Gauge R&R

Problem	Cause	Remedy
Excessive bias	Out of calibration	Calibrate more often
	Improper storage	Store in clean, dry, stable environment
	Improper handling	Train operators
Poor linearity	Gauge wear	Replace gauge
Poor stability	Improper storage	Store in clean, dry, stable environment
	Improper handling	Train operators
Poor repeatability	Improper technique	Train operators
	Gauge wear	Replace gauge or use different type
Poor reproducibility	Operators using different techniques	Train operators

Even attribute (go/no-go)-type gauges can have repeatability and reproducibility issues. If two or more operators use the same attribute gauge with slightly different techniques, the results can be different. Operator controllable variations like the pressure applied by the operator's hands, the angle of view, and the care with which they handle and use the gauge can all affect the outcome. While it is true that you cannot calculate the percent of the tolerance that is used by attribute gauging, you can still determine if different operators will get the same results with the same gauge. Do this by having them both check the same group of parts using the same gauge without seeing each other's results. A sample of at least 20 pieces, at least one of which is defective, is needed for this. A 50- or 100-piece sample with several defectives is better. If they do not obtain exactly the same results by identifying the exact same sample parts as being defective, the gauge is not repeatable or reproducible. Therefore it must not be used for product acceptance because good product can be rejected or bad product accepted by the attribute gauging system, just as with variable-type gauging.

Whether attribute- or variable-type gauging is used, training and, if necessary, practice using the gauge can make a person's measurements more consistent. They may even reduce the operator-to-operator differences. If training and practice do not significantly improve a poor GRR, it may be worth considering a different way to measure the characteristic, one that is more repeatable, more stable, or less prone to operator influence.

DESIGN VERIFICATION AND VALIDATION ACTIVITIES

Industry has long known that solving all your problems up front, before going into production, results in fewer nonconformances being manufactured, less process downtime, and less wasted time and materials. This is the real reason behind doing the preproduction parts approval process (PPAP) and advance product (or process) quality planning (APQP). An important step in either of these is design verification and validation.

Design verification is simply verifying that the design, as documented, meets all the applicable specifications and customer requirements as stated and as intended by the customer, *prior to production*. To verify that the product, as designed, will do what the customer intended it to do is also a part of design verification. Other aspects of design verification may include material reviews, class of fit when parts are assembled, tolerance stack-ups, and print checking.

Well-done design verification is a design review that will identify any flaws or inconsistencies in the design as well as comparing the design to all applicable customer requirements, expressed or implied. If no flaws, inconsistencies, or requirement gaps are found, then the design is verified as meeting the intent and letter of all the applicable requirements. If any flaws, inconsistencies, or requirement gaps in the design documentation are revealed after careful review, they are corrected prior to starting production.

Design validation occurs when the product is actually built, whether a first piece or a number of process samples, and then confirmed by means of inspection and testing that the product actually does meet the design intent in appearance, construction, and performance. If no defects are found, then the design and the manufacturing process have then been validated. Performing a pilot run or a PPAP run of a sufficient number of pieces validates not only that the process and product can meet the design intent, but that they will do so consistently at the desired level of quality. Effectively, a PPAP can be used to validate both product design and the manufacturing process. A completely and properly executed PPAP can also identify any product or process defects that will occur during production, thereby giving you the opportunity to apply corrective actions that will prevent those quality issues during production. If the level of quality observed during the PPAP is lower than desired, then the quality will remain a productivity-hindering issue until it is improved.

EFFECTS OF THE FACILITY

Perhaps one of the most limiting factors in doing any kind of improvement is the plant facility itself. There are different ways that the plant facility limits productivity. The most common are limitations of electrical power, water pressure, and available space. Productivity is also affected by the unevenness of temperature and humidity inside the building. This is especially true in buildings that have been added onto many times over the years. Other effects to be considered are limitations of the amount of weight that the floor can support and the number of people that can be in a given room for the whole workday.

Physical geography, local topography, available space, available electrical power, and available water pressure are all limitations that are difficult and expensive to overcome, if indeed they can be overcome at all. Add to these such things as building codes, zoning regulations, and other local ordinances, not to mention building age and materials, and you have a collection of obstacles toward any kind of plant improvement.

Plant modifications can be expensive, difficult, and time consuming. However, they should not always be ruled out as a means of improving productivity. Judge these on a case-by-case basis, and try to find creative solutions to overcome the limitations. While it is true that in some cases plant expansion is not an option, this does not necessarily rule out all plant modifications in every case. Sometimes there are things you can do to your physical building that can improve productivity. It all depends on your situation.

One example is that the need for more space can be met by building a mezzanine in a room with a high ceiling rather than adding on to the building size. Another example is that the need for more water pressure can be met by a large storage tank that is filled up during off-hours and then pumped out at higher pressure when needed.

EFFECTS OF PREVENTIVE MAINTENANCE

Most people can easily see that preventive maintenance can be good for productivity because it prevents major problems like equipment failures that result in extensive and nonproductive downtime. Preventive maintenance is a two-edged sword. It can be a help and a hindrance, an asset and a liability. When done too frequently or ineffectively, it causes unnecessary downtime. If not done frequently enough, it can make the tooling and equipment worse and more prone to failure, or even result in major equipment failure and extensive downtime.

Equipment must not only have sufficient reliability, but it must also have good maintainability and high availability. A little knowledge of basic reliability can be of real value here. Knowing which components have the worst reliability can help plan the preventive maintenance schedule to minimize downtime and maximize the reparability. Owner's manuals, maintenance history, and engineering knowledge can also contribute to minimizing the downtime of repairing and thus improve productivity.

When the maintenance department schedules maintenance activities in an area, they need to consider how that area is going to be affected by the maintenance activity and try to minimize their impact on production. If an area is busier at specific times or seasons, schedule the activities around them. Look at personnel and material movements, as well as traffic patterns in the area to be maintained, and plan activities accordingly. Check with production control or manufacturing about production plans for the area, and schedule accordingly. Anything maintenance can do to minimize their impact on production will prevent them from unnecessarily decreasing productivity.

Typically, the maintenance activities are grouped into specific time periods to be called upon at their fixed time intervals. To minimize downtime, you must minimize maintenance time. This can be done by creating a small inventory of preventive maintenance kits, each kit containing all that is needed for the preventive maintenance for that particular maintenance interval. Then when the maintenance interval is up, the maintenance employee simply takes the kit and does what it calls for without having to waste time looking for tools or materials because they are already there in the kit.

Some of these cross-functional productivity-improving activities are discussed in greater detail in Chapters 5, 7, 8, and 9 of this book.

4

Productivity and Human Resources

EMPLOYEE ORIENTATION

No job function is so isolated as to be totally independent and unrelated to other job functions. All jobs affect others and all are affected by others. Procedures in ISO 9001–based management systems are extensively integrated and interconnected if they are properly designed and well implemented. Therefore, the orientation of new employees must also be broad enough to show the new employee where and how their particular job is interconnected with others and how it fits into the larger picture. This information can help new employees work and make their decisions more wisely and beneficially.

Employees who think their job is trivial or that no one is affected by their output may be less careful or conscientious about their work. They are also more likely to be less productive or more error prone, whereas employees who think their output is important are likely to be more careful about quality and productivity. Those who understand how their job fits into the whole operating system of the company will more easily understand the importance and effect of their output and therefore may be more productive. A broader, more integrated and comprehensive employee orientation program for new hires may help new employees to understand how their particular job is integrated and interconnected with the rest of the company.

To help a new employee see how their particular job affects productivity elsewhere in the production process, many companies give new employees a detailed tour of the production process with explanations of how their responsibilities affect productivity. It is not unusual for a new employee to meet with a variety of department representatives that walk through the process with the new employee and give their particular perspective on things. This has the added advantage of the new employee meeting a variety of people in the company and learning who is responsible for

Human Resources

Policies: ☐ Smoking   ☐ Dress code   ☐ Phone usage
☐ Computer usage
☐ Security   ☐ Nondisclosure   ☐ Lunch & breaks
☐ Attendance

Procedures: ☐ Parking   ☐ Evaluations   ☐ Emergencies
☐ Clocking in & out   ☐ Training records

Human Resources signature_____

Quality

☐ Introduction to quality system   ☐ Work instruction usage
☐ Record keeping & locations   ☐ Workmanship standards
☐ Quality policy   ☐ Company goals and objectives

Quality signature_____

Manufacturing

☐ Plant tour   ☐ Process flow   ☐ Material flow
☐ Departmental organization chart   ☐ Companywide organization chart

Manufacturing signature_____

Other Departmental Orientations

_____Signature_____
_____Signature_____
_____Signature_____

FIGURE 4.1
Example of new employee orientation checklist.

what. Often there is a checklist that accompanies the new employee on which each department signs off their particular part of the orientation. An example of an employee orientation checklist is shown in Figure 4.1.

POLICIES AND PROCEDURES

Another thing that influences productivity is the company's own policies and procedures, habits, and paradigms. These can reduce productivity by simply causing things to be done poorly or in a way that is counterproductive. This can be due to a lack of operator training, insufficient management expertise, old habits, poor equipment, old technology, or even a lack of resources. Hence, manufacturing must make its needs known and use what is has most wisely. Nevertheless, there are certain modes of operating that are inherently counterproductive. These counterproductive policies and traditions are often the result of preconceived notions and bad habits.

Policies and procedures that are counterproductive may be either poorly thought out or simply one person's opinion resulting from their own perception and experience, rather than taking into account the bigger picture. It could also be a habit from someone's previous work experience that does not fit well into the current company.

Inadequacies of policies and procedures are identifiable by how well they work, how they are enforced, and whether or not they accomplish their intended goal.

Activities that decrease productivity can occur in any department, but in certain departments, nonproductive or even counterproductive activities can occur. In manufacturing departments, the choice of trainers, calibration, labeling of work in process, and tooling issues are the most common.

Human resource departments are not immune from having an effect on productivity, good or bad. Inadequate training, hiring employees with insufficient literacy levels, providing limited or poor-quality orientation, and improper or ineffective employee evaluations are all ways in which human resources can cause productivity to be less than it otherwise could be.

TRAINING

As discussed in Chapter 3, the choice of trainers, the training technique and effectiveness, and the overall training plan can affect productivity. But these are not the only training issues that have an impact on productivity. Other productivity-related training issues are

- Scheduling and duration of the training
- Applicability of the training to what the employee actually does
- Verification of training usage on the job
- Consequences of not applying the training to the job

There is often a need to train people away from the workstation. Such training can be for varying lengths of time, from a matter of minutes to a week or two. If practical, off-line training should be done before or after production time. Time spent on such training programs is time spent away from producing product. Scheduling such training during off-hours may require the payment of overtime to employees and may even be a hardship to some employees. Nevertheless, time spent in training is time spent out of production, so if scheduling training is practical before or after normal shift hours, it is worth considering for the sake of productivity.

If this is not practical, another approach can be considered. Perhaps breaking the training up into several shorter training sessions held during normal work hours may be more feasible. Alternatively, you can train one employee at a time, while the others pick up the slack much as they would if the employee was out sick. Some advantages to multiple short training sessions are that it makes it easier to absorb the material and gives the employee time to digest the new knowledge. It may also provide an opportunity for the employee to more immediately put into practice what they have learned. Since their learning is more gradual, their implementation of the new knowledge will be also. This may make for better implementation of the training.

Training that is not applicable to an employee's occupational activities is soon forgotten, so the time spent on it is wasted and nonproductive. It is the *employee's perception* and judgment of applicability that affects knowledge retention and use, *not management's assessment* of applicability. Employees must be able to clearly see how the training affects them and their particular job. The more immediate and measurable the impact, the better it is. Employees will regard training they consider to be nonapplicable as a waste of time and effort, regardless of whether the nonapplicability is real. It is only their perception that matters. Employee communication, input from various stages in assembly, and employee exposure to innovation can all help management to determine what training will be perceived as really applicable and what will not.

Verifying that the training is actually being used on the job is not usually accomplished effectively by a checklist or a line on which to place one's initials. These are usually merely regarded as additional paperwork activities, and employees do not consider them productive. They can be filled in immediately before an audit, rather than be evidence of actual performance of the task. Seeing a check in a box proves only that someone put a check in the box. It does not prove that something was actually done.

System audits are also not an effective way to verify the implementation of training because they are usually too infrequent to effectively monitor the training implementation. Likewise, video cameras are not a good way to verify implementation of training because they can make employees feel untrusted or worsen labor relations. This only hurts productivity and never improves it.

Effective implementation is best verified by the employee's peers. Having them check each other not only verifies implementation, but also enables them to help each other with knowledge gaps, and even cooperatively figure out the best way to implement what they have learned. Verification of implementation can also be effectively done by the trainer or supervisor, who can monitor and check the employees. This monitoring is initially done frequently and can be gradually decreased in frequency if it shows positive results. When it seems prudent, the monitoring may be gradually discontinued. One advantage of having trainers or supervisors verify implementation is that they can help solve any problems with implementation if and when they occur. By monitoring they may see ways to improve the training.

Failing to apply training to the applicable operations is an employee discipline issue, but it can also be an issue of training effectiveness or misconstrued applicability. Conceivably, poorly selected or presented training and nonapplicable training could have little or no effect on productivity. In any case, an employee's not applying the training to the job is something that needs to be investigated and addressed. Reassess the effectiveness and applicability, and then find out why the employee is not implementing the training. Finally, take appropriate action.

In the training of employees, human resources has several roles. They need to create and enforce policies that are training-friendly and facilitate training for productive employment. They should cooperate with the training personnel and manufacturing departments with scheduling times and places for training. They may even verify that the technical level of the training is appropriate for the targeted employees. This is especially important with home-grown training programs developed by engineers, which may be too technical or provide too much information too quickly.

COMPENSATION AND LITERACY LEVELS

Hiring practices of the human resources departments can also sometimes have an effect on productivity. Of course selection of prospective employees for hire requires matching the education, intelligence, and skill levels of the candidate to the requirements of the job. Equally obvious is that underestimating the job requirements or overestimating the abilities of the candidate can result in a low-productivity employee. However, these are not the only considerations. The availability of the prospective employee or the wage they are willing to accept can, and sometimes does, bias this matching of the employee to the job. An employee willing to work for a lower pay may sometimes actually cost more when you consider the cost of their lower productivity if they turn out to be a lesser qualified employee. Some people will say this is rare—that it is the exception rather than the rule. That may be so, but every time this is the case, it causes productivity to be reduced.

A lower literacy level in employees may or may not also have an effect on productivity. Less-literate employees may sometimes take longer to train. When this is the case, the increased learning curve delays the time it takes for the employee to get fully up to speed and reach their maximum productivity. This is not to suggest that employees of lower literacy are less trainable or have less potential, but rather, that the training may sometimes take a little longer. The longer the time required to get up to speed, the more the productivity is affected.

On the other hand, even employees of less-than-average literacy may have such talent and aptitude as to be an asset. They may effectively improve productivity. This depends mainly upon their creativity, learning ability, adaptability, motivation, energy levels, and general health, as well as their passion for their work. These characteristics can make all employees more productive. Maximum productivity is achieved when good education, intelligence, and skill levels are combined with creativity, learning ability, adaptability, motivation, passion for work, and so forth. Such employees may or may not desire a higher starting wage, but the increased productivity makes them an asset to the company.

WORKLOAD

Workload can either help or hinder productivity. As mentioned in Chapter 2, merely having fewer people do more work is not necessarily going to result in productivity improvement. While trimming away excesses in the workforce can result in increased output per labor-hour, the improvement is neither linear nor always positive. For any workforce there is a maximum amount of work that can be productively done per person, as well as by the whole group. To put it another way, for every given workload there is a minimum number of people it will take to accomplish the work productively. When the number of people working falls below this minimum, productivity will begin drop off due to increased errors and nonconformances, poorly implemented corrective and preventive actions, and increased malpractice.

Maximum workload per person is variable. Group performance for a given workload is less variable and easier to measure. It is not really hard to tell if the maximum workload for best productivity has been exceeded. There are clear signs if you know what to look for. The most common and immediate signs that the workload is too great are an increase in workplace accidents or safety violations, increased absenteeism, increased clutter, and disorderliness of the work area. Other signs are an increase in shortcuts or cutting corners in procedures and an increase in audit findings. Still another sign of employee overload is an increase in the proportion of nonconforming product being produced. All of these symptoms of excess workload will decrease productivity. Latent signs of excessive workload are an increase in the proportion of customer complaints and an overall decrease in satisfaction by both supervision and employees. This may show up at employee evaluation reviews.

The human resources department can have an effect here, too. By giving information to manufacturing management concerning workplace accidents, absenteeism, employee evaluations, and other symptoms of work overload, they can help management identify the signs of employee overload. They can also give support to manufacturing's efforts to have the right number of staff to prevent or remedy the overload and, in doing so, improve productivity.

EMPLOYEE EVALUATION

Many companies perform employee evaluations periodically, often annually. These evaluations, often called reviews, are regarded as a way to measure how well an employee meets the company's expectations, with the implication being that meeting the expectations is synonymous with meeting the company's needs. Increases in the employee's compensation are often tied to this evaluation so to act as a motivator for the employee to perform better. Discussions with the employee about their particular job expectations and performance are looked upon by management as a type of job counseling.

The question is this: How does the employee's job performance truly relate to productivity? More specifically, does a poor evaluation really mean the employee is less productive than an employee with a better evaluation?

These questions open the door to the debate on whether or not an employee's evaluation should or should not be tied to productivity. If so, should the evaluation be entirely about productivity or should productivity be only one factor among several? Should productivity be more of a factor in some job evaluations and less in others, depending on the employee's specific responsibilities?

The answer is that the productivity of an employee should be considered in an employee's evaluation *only if their productivity is something the employee has real control over* and to the extent that it affects the productivity of the company as a whole. Constantly being held responsible for and evaluated on something the employee has no real control over not only frustrates the employee and management, but it can supply motivation for the employee to resign and seek employment elsewhere. When evaluating an employee, consider the extent to which their particular occupational responsibilities truly affect the company's productivity. Then be sure to consider how much real control the employee actually has over their productivity. Let these two things determine the extent to which productivity will be a factor in the employee's evaluation.

It is not unusual for management to overestimate how much real control an employee has over their productivity. It is very much a matter of how much control the employees have over the manufacturing process. For employees to truly have control over their own output, three conditions must be met:

1. They must know what the output is supposed to be.
2. They must have a means of accurately comparing the actual output to what it is supposed to be.
3. They must have a means of adjusting the output to reconcile any difference between what the output should be and what it really is.

If the employee lacks any one of these, or if they have any one of them only partially, then the employee does not have real control over their own output. Hence, they have no real control over their productivity. Evaluating an employee on something they have no control over does not improve productivity, but it does build resentment, disloyalty, and the desire to leave the company.

Although employee evaluations can be motivational, there are times and circumstances where they actually reduce motivation. Any time a company reduces the motivation of their employees, they are risking a decrease in their own productivity.

Employees are motivated to work for companies that they like and respect. This applies to individual supervisors as well. Employees will put more effort into doing a good job for companies and supervisors they like and respect, compared with those they don't. The difference in job performance is not necessarily intentional, although for an individual employee it might be. So anything that management does to cause the employee to not like and respect the company or an individual supervisor will reduce motivation. Productivity will be affected by this to the extent that productivity is a part of their job performance and they have control over it.

It is human nature that we respect those who respect us and have little or no respect for those who have little or no respect for us. Certain actions related to employee evaluations can cause employees to reduce their respect for the company or supervisor. One of them is dishonesty from management. Another is raising hopes with no intention of fulfilling them, which itself is a form of dishonesty. Dishonesty can also occur in other forms like false promises, inaccurate reviews, and management not following their own polices or enforcing them differently for different employees. Such behavior by management, and especially human resources, can destroy an employee's respect for the company or their supervisor. That loss of respect reduces employee productivity and therefore can affect the productivity of the whole company.

Evaluation reviews that are very late send the message to the employee that the employee's review is not important to the management. Late reviews tell the employee that their review is not a priority. The later the review is, the lower the priority that the employee's perceives the review to be in the eyes of management. True or not, this is the employee's perception. This may not actually be the case but it is the message sent if management does not give it enough priority to enable the review to be given to the employee on time. If management does not seem to care enough about the evaluation to complete it on time, how is the employee supposed to take it seriously? If the employee does not take it seriously, then it will have much less effect on their performance, and that can mean the review has less effect on productivity.

Check each box when an orientation item is completed, and then sign off. When all items are completed, return the checklist to the Human Resources Department.

5

Productivity and Your Quality Management System

PDCA CYCLE

The year 2000 and later revisions of the ISO 9001 standard for quality management systems, and its variants, along with other manufacturing operating systems, is based on the Plan–Do–Check–Act cycle, also known as the PDCA cycle or the Deming cycle. This cycle is a repeating process of performing the four basic actions that a company performs to produce their product. These actions are often referred to as sections, stages, or phases of the cycle. See Figure 5.1 for an illustration of the PDCA cycle.

The four sections in the cycle are plan, do, check, and act: Plan how you are going to fulfill your customer's requirements. Do the activities you have planned. Check that the results of what you have done are in accordance with the plan and give the expected results. Act on any activity or result that is at variance with the plan or with the customer's requirements, so as to reconcile what you actually did with the original plan and intended results.

If any one or more of these stages are not done properly or done out of order, poor productivity can result. Each stage must be completed thoroughly. Starting the next stage too soon, before the previous stage is sufficiently completed, will cause preventable problems to interfere with productivity. Besides this, gaps and inconsistencies in the cycle are themselves culprits that can result in poor productivity.

Developing the *plan* is the first stage of the PDCA cycle. The planning stage need not be entirely redone for each customer contract because much of the planning is common to all customers and all products. These particular aspects of planning can be done once, remain permanent, and applied to other contracts. Examples of this are company organization and apportionment of management responsibilities, the development and

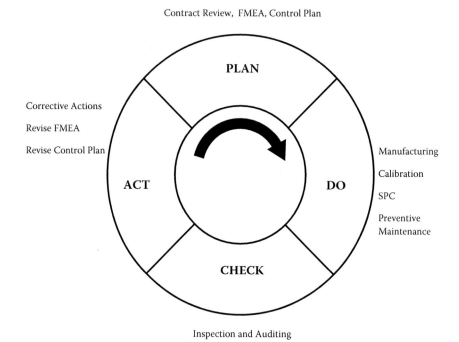

Contract Review, FMEA, Control Plan

PLAN

Corrective Actions

Revise FMEA

Revise Control Plan

Manufacturing

Calibration

ACT

DO

SPC

Preventive
Maintenance

CHECK

Inspection and Auditing

FIGURE 5.1
PDCA cycle.

implementation of the quality system itself, physical layout of workstations, identification and approval of suppliers, hiring practices, most work instructions, development of some training programs, and so forth.

Other planning activities may be specific to a particular product model or customer order. These will include work orders, routing sheets, specific operation training programs, some work instructions, and even test procedures.

It is during the planning stage that contract review is most important. There are several reasons for doing an interdisciplinary contract review:

- The management of each department knows best the capabilities, methods, and issues of their particular department.
- It enables issues to be resolved before production on the order actually begins, so that time and other resources are not wasted solving the issues during production.
- It provides the departmental managers a heads up on what is in the pipeline so that the department can adequately prepare.

- It facilitates cooperation and communication between departments, which itself can benefit productivity.

To properly plan is really about preparing for what is coming so as to maximize productivity and prevent problems that can reduce productivity. Such preparations may include but are not limited to specialized training, design and development of new tools or fixtures, purchasing or otherwise making available additional equipment, rescheduling planned repairs, determining inspection methods, changing inspection plans, and the like. In the case of a small single-owner company, even the owner or a salesperson alone doing contract review would not necessarily have the intimate knowledge of each department that the department managers would have. Consequently, a contract review done by the individual would not be as thorough as an interdisciplinary contract review. That lack of thoroughness could cause issues to go unnoticed until the department actually begins work on the order. Then it is too late. A contract review performed by a single individual does not allow the other department managers to plan for what is coming as well as they otherwise could. Having to solve problems on the fly diverts attention and resources from manufacturing, thereby slowing it down. The effort expended on solving problems reduces productivity, whereas a more thorough and interdisciplinary contract review would have brought the issue to light before the work began, thereby enabling the problem to be dealt with before production begins on the order. Then the problem would have much less of an undesirable impact on productivity, or even no impact at all. Even if the problem could not be solved, it might be compensated for, or at least potential resolutions might have started earlier. This, too, would reduce the negative impact on productivity.

Do not overlook the advantage of planning workload distribution, staffing, and cross-training of employees of particular work cells. These can improve productivity. If the manager of the department is involved in the contract review, he or she will know ahead of time what is coming and best plan how to portion out the work and the cross-training that is needed.

Planning is not just a matter of contract review. It should include a process failure mode effects analysis (PFMEA). This document, when properly made and used, is one of the most effective planning tools. The manufacturing process must already have been planned in detail in order to do this analysis. A flowchart or any process flow document will help you to include all the process steps in the PFMEA. For a good instruction on how to make a PFMEA, consult one or more of the suggested readings found at the end of this book. The PFMEA identifies all possible ways the manufacturing process can fail and requires that preventive action be taken *before production begins* in order to prevent failures. The failures acted upon are selected as the ones having significance at or above an action threshold on the PFMEA, but it does not forbid taking action on failure modes that are less probable or less severe.

To create the PFMEA, first list all the process steps. Then for each process step, list all the ways the step can fail or in any way produce a defect. These are the failure modes. Then for each failure mode, estimate the probability, severity, and detectability by the customer on a scale of 1 to 10. These are multiplied by each other to calculate the risk priority number (RPN) for each failure mode. Then a threshold RPN value is determined, above which preventive action must be taken. This is typically an RPN of 100. Information on making and using PFMEAs is available in several publications. You can see some of them in the Recommended Readings list. An example of a PFMEA is shown in Figure 5.2.

From the PFMEA you make the control plan. The control plan tells show each step in the process is controlled. Included in the control plan is the measurement system description. Control plans tell the operator what to do to contain defects and prevent their spread. It also gives a corrective/preventive action to be applied until control is restored. PFMEAs and control plans must be updated during production and after production using information from defect and customer return analysis.

It may be that for a particular characteristic you may choose to use statistical process control (SPC) as a control method. Use the PFMEA to determine what parts and characteristics to do SPC on, and then decide exactly which SPC technique to use; for example, an X-bar and R chart with a subgroup size of five pieces checked twice a shift, and monthly Cpk analysis. Then document exactly what the SPC technique and sampling will be on the control plan. Do not forget to include on the control plan what to do when the chart shows an out-of-control condition.

Properly planned and implemented SPC can be an effective control method to use, but do not neglect other methods. First piece approval, in-process inspection or testing, process audits, and some other methods have merit as well. Quality engineers are especially adept at choosing the control method. Different characteristics may be controlled differently,

	Process Failure Mode Effect Analysis							
Part Number: 1234567-2				**Part Name: Type B widget**			**Drawing Number:**	1234567
Process Function Requirements	**Potential Failure Mode**	**Potential Effects of Failure**	**Severity**	**Potential Cause(s)/ Mechanisms of Failure**	**Occurrence**	**Current Process Controls**	**Detectability**	**RPN**	**Recommended Actions**
1. Material procurement	Wrong material	Product failure	10	Supplier did not ship to specification	1	Spec on purchase order	7	70	No additional action required at this time
			5	Missing or wrong material certification	2	Cert checked at receiving inspection	1	10	No additional action required at this time
2. Drill mounting holes	Holes missing	Inability to assemble	10	Skipped operation	1	Operator use control plan and sign off on routing sheet	5	50	No additional action required at this time
	Wrong hole size	Bolt will not fit	9	Wrong drill size	2	Tool check at setup and 1st piece inspection	5	90	No additional action required at this time
	Wrong hole position	Inability to assemble	10	CNC index error	2	Check at setup and operator training	4	80	No additional action required at this time

(Continued)

FIGURE 5.2
Process failure mode effect analysis.

Process Failure Mode Effect Analysis

Part Number:	1234567-2			Part Name: Type B widget			Drawing Number:		1234567
Process Function Requirements	**Potential Failure Mode**	**Potential Effects of Failure**	**Severity**	**Potential Cause(s)/ Mechanisms of Failure**	**Occurrence**	**Current Process Controls**	**Detectability**	**RPN**	**Recommended Actions**
			10	CNC program error	2	Program approve at 1st piece inspection	5	100	Correct program to be verified by 1st piece inspector
3. Move to assembly department	Brought to wrong department	Delay in manufacturing	5	Improper labeling or lack of material handler training	6	Training of material handler and part label	4	120	Add destination label and verify routing is with part
	Damaged in transit	Part is scrap	10	Part not protected during transit	6	None	3	180	Wrap in bubble pack then put in tote box

FIGURE 5.2 (*Continued*).

and the control plan must tell the operator which control method will be applied to which characteristics.

The *do* portion of the PDCA cycle is a matter of working according to the plan. Not following the plan or following it poorly is what reduces productivity in this stage. Each department's part of the plan must be well known and understood by the supervisors and employees in that department. The resources necessary to carry out their task must be available before production begins, and employees must know how and when to access them.

Test runs, pilot runs, or a full-blown preproduction parts approval process (PPAP) can identify issues right at the end of the *plan* stage or the beginning of the *do* stage of the PDCA cycle. If no PPAP is required, it might be wise to have a process rehearsal, dry run, or pilot lot. This could reveal any gaps in training or even show the need to modify the layout of an assembly line or work cell. It may also provide information indicating that a revision to the PFMEA or the control plan is necessary. Input from the employees doing the labor is also of value here.

The PPAP is a thorough and standardized way to determine if the manufacturing process actually can produce parts with a low enough defect rate to be productive. It also proves to the manufacturer and to the customer that the process is controlled well enough so that quality issues will not impede production. When performed and used properly, PPAPs enable the manufacturer to identify and address design, manufacturing, and quality issues before the actual production run begins. This is so that design, manufacturing, and quality issues will not impede productivity.

Some companies mistakenly use the first production run for their PPAP samples or do the PPAP concurrently with the first production run. Doing either of these is a mistake because it defeats the purpose of the PPAP. Another mistake that defeats the purpose of the PPAP is not manufacturing the PPAP samples using production tooling, methods, materials, or people. The purpose of the PPAP is to verify that the actual manufacturing process used in production can and will produce parts with the level of quality expected by the customer and to give the manufacturer an opportunity to solve all quality and production issues before production begins so as to have smoothly operating, highly productive, high-quality production process. To verify the process is good and deal with the quality problems beforehand, the PPAP must be completed using actual production process. This requires using the actual equipment, methods, materials, and people *before production of the order begins.* Not using actual production tooling, methods, materials, and people will not give PPAP results that truly represent the actual manufacturing process that will be used in production. Therefore, it will not be a valid process verification, nor will it be indicative of quality problems you will encounter during actual production.

Check, the third stage of the PDCA cycle, is to examine what was done and compare it to the plan. Production results also need to be checked in relation to the customer's requirements. Since true productivity is measured only with good product, it is important to distinguish which products conform to the customer's requirements and which ones do not. Checking is what tells you the defect rate (among other things) that affects productivity. Methods for checking always include auditing, inspection, and testing, but they may also include discussions with employees or the customer. These may be done informally immediately following the completion of a task, or more formally at a planned meeting. The management review process required by the ISO 9001 standard and its variants is also a checking activity.

Auditing is usually employed to check how well, that is, how thoroughly and intelligently, the process planning was accomplished; how well it meets the manufacturer's and customer's needs; and how well the process was implemented. Auditing checks the process, whereas inspection and testing check only the product itself, rather than the process. Inspection and testing will tell you if the product, as it was manufactured, actually meets the plan and meets the customer's requirements. Discussions with employees can also reveal how well the planning and doing stages were accomplished and bring to light issues that may not be picked up by inspection, testing, or even auditing. Management review meetings are necessarily broader in scope, but they can also include a review of major projects and customer contracts. The management review meeting can help all the managers understand what happened, why, and how. Management review meetings can be a good forum for addressing systemic issues.

Finally, *act* is the stage of the PDCA cycle that is most crucial to productivity improvement. No matter how they are identified, variances from the plan or from the customer's requirements need to be acted upon. To act on these variances is the fourth part of the cycle. Closing the gaps, dealing with lessons learned, and making improvements identified earlier in the cycle are all ways that productivity can be improved in this stage of the PDCA cycle. Well-thought-out and implemented corrective and preventive actions are a valuable tool to use for increasing productivity.

Do not forget to follow up with actions like training, resource reallocation, and verification of implementation so that corrective and preventive actions do not become merely paperwork exercises but instead can result in real productivity improvement. After the actions are implemented and verified, the PDCA cycle then begins anew with more planning on the next contract, with lessons learned being applied. This is when the PFMEA must be updated to include the lessons learned and make the severity and probability assessments more accurate. Many companies make the mistake of not updating the PFMEAs and control plan. These are living documents designed to aid in the planning stage, so that productivity-reducing issues can be prevented. PFMEAs and control plans are highly effective when properly used and kept up to date.

QUALITY MANAGEMENT SYSTEM ISSUES

Sometimes systemic issues interfere with productivity, thereby making it less than it would otherwise be. By *systemic issues* is meant the characteristics and situations that occur in the company's own operating system. Every company has a system, whether it is a formally documented quality management system or an undocumented operations system. A system is a collection of related procedures by which the company carries out the daily activities of its business. ISO 9001, while often called a quality system, is effectively an organizational management system. In any case, whatever system you have, regardless of how well or how poorly it is documented, affects your productivity. Characteristics of the system, and the situations the system causes, affect productivity for better or for worse. There are several ways in which the effect can be detrimental to productivity.

Documentation conflicts are one issue that affects productivity in an undesirable way. Giving conflicting information to employees not only causes confusion and errors, but it also calls for a resolution of the conflict. Only then can it be determined what the operator is supposed to do. This can use up time, divert people's attention, divert resources, and interrupt the flow of materials. The effect on productivity occurs as a result of the conflicting information in the documents and can happen in different ways—alone, in combination, or all at once. An example will help to explain.

> Employees get work instructions from different sources. Typically this can be shop routing paperwork, operation sheets, prints, memos, and so forth. What if there is a conflict on what the operator is supposed to do? A conflict could be something like an operation sheet or routing form telling the operator to fasten two parts together using a #10-32 bolt, but the print says to use a #10-24 bolt. This could interfere with the manufacturing flow in several ways:
>
> - Time is wasted while the person who can determine which bolt to use is tracked down, comes to see the issue, and makes a decision.
> - The operator looks at the operation instructions and uses the #10-32 bolt the instructions require, only to have inspection reject the part because it is not according to the print, which requires the #10-24 bolt.
> - An in-house customer representative sees the issue and requires that work in process and stock be pulled and placed on hold until the issue is resolved.

These are just three possible scenarios. Any one of these alone could reduce productivity and can even bring production to a halt. More than one of these could happen, wasting even more time, manpower, and other resources. Finding and correcting these conflicts ahead of time prevents them from reducing productivity.

Documentation loops also impede productivity. One kind of documentation loop is when system documents reference each other in a circular manner without really telling you what you need to know. Here is an example:

> You are at a point in the manufacturing process where you test the product. The work instruction says to connect the test cable to the connector and test for the amount of time and at the temperature determined by the print requirements. The print tells you to test product per procedure QAT-111999. You then look at the test procedure QAT-111999 and it says to test for the amount of time and temperature as shown in the work instructions, so you are back where you started and still do not have the information on the test time and temperature that you need.

Document loops are not usually that simple or that obvious, but you get the idea. The person doing the test has followed the document trail but still does not know how long the test should be or at what temperature.

Document loops can affect productivity in the same ways that document conflicts can. They need not occur among different documents; they can be within the same document. In any case, the effect is the same as a document conflict, and so are the solutions. Careful proofreading by someone other than the writer of the document can help identify conflicts and loops. Carefully performed auditing can also find them. Like conflicting information, finding and resolving document loops before production begins can improve productivity by not wasting time and other resources and not interfering with the flow of material through the process.

Document incomprehensibility is a more common issue. Usually documents like work instructions and routing sheets are written by engineers, often manufacturing engineers or even quality engineers. They are written in a way that makes perfect sense to the writer but may be incomprehensible to the user of the document. Vocabulary, sentence length and complexity, and logical progression of thought are all factors in comprehension. The assumptions that the writer makes about the operator's knowledge and thought processes when using the documents can also be a factor in document comprehensibility. English as a second language is another factor in the employee's ability to understand and properly use the document. Differences in age, operator education and experience, and even social circles are factors as well.

Documents have to be written in a manner that is easily understandable *to the user*, not to the writer of the document. Before a document is approved, the writer, the approver, and the person using the document should all get the same understanding just by reading the document. If any explanation at all is necessary for that common understanding to happen, the document needs to be revised to include that explanation, no matter how obvious it seems. The fact that an explanation was even necessary is proof the document comprehensibility was inadequate as originally written.

In addition to document conflicts, loops, and incomprehensibility, another problem is the specificity of the document. If a document is either too generic or too specific, it can cause problems. Documentation specificity in work instructions, procedures, and the like are supposed to be a means to ensure consistency. This consistency prevents process variation and defects, both of which decrease productivity. The documentation specifies who, what, and how in order to prevent variation. That is how productivity is maintained by specifying activities on documents.

The theory goes like this: If a process is a good and productive one, running that process consistently keeps your quality and productivity consistent. Having your process steps documented very specifically creates a standardized and therefore consistent way to perform your manufacturing operations. Consistency is maintained by the operators' following a single standardized method of performing the operation. This consistency will prevent unwanted variation and defects. Both variation in the process and defective production decrease productivity, so preventing them is good for productivity. This is the theory, but the reality does not always match it.

This way of doing things requires that every time the process is revised, all the appropriate paperwork must also be revised and the operators trained to the new revision. If that is what actually happens in your company, you are on the road to consistency and are reaping its benefits. However, keeping up with documentation changes is not the only problem.

Documents that are too specific can also create problems and hurt productivity. The more specific an instructional document is about a process, the more optimized the process must already be. Process steps and descriptions that are not optimal but nevertheless are very specific can prevent the optimal process from ever being developed. Also, operators differ in manual dexterity, strength, left- and right-handedness, height (and therefore angle of vision), and motor skills. They even vary in the way their brains process size, shape, and motion. For any operator to work at peak efficiency and therefore maximum productivity, the written procedure must leave room for these differences.

On the other hand, documents that are too generic do not give the operators enough information to be consistent with the intent, but they do give operators a lot of room for variation. A work instruction that is too generic provides an opportunity for various operators to do things differently, or for the same operator to vary what they do. Ad-libbing and customizing the operation, whether intentional or not, is more easily allowed by instructions that are not specific enough. Consistent performance of an operation and the productivity benefits of that consistency are compromised by excessively generic work instructions, operation sheets, routing sheets, and other documents that tell your operators what to do.

Neither excessive freedom from too generic an instruction, nor insufficient freedom in an instruction that is too specific, is good for productivity. The goal is to have the right amount of specificity, neither too much freedom nor not enough. The exact amount will vary with the design, technology, ergonomic situation, training, and degree of differences between operators. Furthermore, different operations will require different amounts of specificity.

Sometimes the operators themselves find a better way to do things, which can improve productivity. They should be given the freedom to do that, or at least feel that they can discuss alternative procedures with the appropriate people. This kind of operator empowerment is becoming more common and has proven to be good for productivity.

Over- or underdocumenting can also impede productivity. Although these terms are sometimes used to describe documents that are too generic or too specific, they more properly apply to the number of documents applicable to an activity or operation and the amount of documentation, rather than to the content or specificity of the documents. A real-life example I once saw in a particular company was overdocumentation for the creation of a purchase order. One procedure was to write a purchase request, which started the development of the formal purchase order. A second procedure told how to create the purchase order, a third document described the purchase order approval, and a fourth document described how to amend an existing purchase order. A fifth document described the distribution of the purchase order. A sixth procedure was for placing the order, and a seventh told where and how to file it. Seven procedures were developed to accomplish what is essentially one task: the purchasing of a commodity. Most of those procedures, if not all, could have been put into one procedure that would have been only three pages long or at most four. The purchasing procedure was overdocumented.

Another way overdocumentation occurs is by documenting something that doesn't need to be documented. ISO 9001 requires that you document only six procedures, plus whatever you need to document to ensure sufficient control over your product realization (manufacturing the product). It does not require that every single thing you do be documented. It is true that the ISO 9001 standard does not forbid you from documenting everything. Nevertheless, documenting more than is necessary can hinder productivity.

So how does overdocumenting affect productivity? One way is through your auditing. The more documentation you have on a specific activity, the longer it takes to audit properly. Auditing manufacturing personnel, especially the operators, slows down or otherwise burdens the manufacturing process. The longer the audit takes, the longer the operator is prevented from working at full capacity. Besides that, more extensively documented procedures require more extensive auditing, provide more opportunity to have nonconformances, and so are more likely to require corrective actions. If the internal auditors are production people, longer audits keep them away from production for a longer period of time.

Underdocumenting can be just a bad for productivity. If a process is underdocumented, then there is not enough description to adequately tell people what to do. The lack of description can also make the process harder to audit. There will be insufficient instruction to achieve the consistency that the documentation is supposed to create. Therefore the benefits of consistency, namely, reduced variation and the manufacture of fewer defects, cannot be realized. Another result of underdocumentation is that like a procedure that is too generic, it allows too much freedom of variation. This creates more opportunities for errors and their associated defects. Naturally these lower productivity.

CALIBRATION ISSUES

Carrying out the calibration system can reduce productivity if the system is poorly planned or poorly implemented. The effects of calibration activities on productivity are more subtle than other activities, except if a measuring device is so inaccurate that it results in nonconforming parts being made or sent to the customer. In that case, the effect of calibration on productivity is obvious.

It often seems as though the hardest thing about maintaining a calibration system is keeping all the calibrations of all the measuring devices up to date. It is not uncommon for even small companies to have from 1 to 5,000 measuring devices on which to keep calibration records. Add to that the time it takes to locate all the measuring devices, calibrate them, and update all the records. Now consider the maintenance and calibration of the calibration standards themselves, the problem of lost gauges, and the time to repair or replace broken or lost ones. This whole process gets repeated at every calibration interval, which can be several times a year for many items. Keeping the whole calibration system up to date is a monumental task. The error here is that many companies make it more difficult than it has to be. Trying to keep accurate records of all the gauges and keep them in calibration is made more difficult by doing more than is necessary and by cumbersome, poorly planned systems.

The most common calibration system errors that decrease productivity are

- Calibration intervals incorrect
- Calibration system with no means to modify the calibration intervals
- Calibration due date too specific
- Calibrating items that do not need calibration
- Calibrating at the wrong place or under the wrong conditions
- Improper calibration record keeping
- Unclear responsibilities for calibration

There are three common strategies that companies use to determine calibration intervals, and all three of them are unwise. One common strategy is to have all gauges of a specific type be due for calibration in the same month. For example, all micrometers are to be calibrated in January, all calipers in February, all height gauges in March, and so forth. This is not a good idea. If they are all due in the same month, but not on the same day, it will be more difficult for the calibration person to keep up so that no micrometer calibration expires before the next one is due. An even worse scenario occurs if all the micrometers are due on the same day of the same month. How does a measurement on a production part requiring a micrometer get accomplished if every micrometer in the company has been recalled for calibration? The measurement may be skipped, or the operator may have to wait for an available micrometer. Either is bad for productivity. Another scenario is that if all the micrometers are due on the same day, but it takes several days to calibrate them all, it is inevitable that some micrometers will have their calibrations expire. This will occur unless the calibration begins several days before it is actually due. Another effect of this is that the probability of any particular micrometer being out of calibration is greater.

It is better to have different types of gauges all due in the same month so there are always some measuring devices of any particular type available that are not due for calibration and which can be used by production. That way there is no waiting for an available device and no skipping of measurements. Therefore calibration will not interfere with production and lower productivity.

A second strategy for determining calibration intervals is to determine the interval only by gauge type. For example, all calipers get calibrated every three months or all volt meters get calibrated semiannually. This, too, is not the best use of calibration resources. Think of this: Why should an 18-inch caliper that gets used once every other year and a few 12-inch calipers that get used just a few days a year get calibrated every three months like all the other calipers that are used every day, just because they are calipers? More than gauge type alone must be considered. You also must consider usage rate. How often a gauge is used is a better indicator of how often it should be calibrated. Usage rate should not be applied to gauge type but to individual gauges and locations. For example, a 12-inch caliper may be used daily in receiving inspection, but other 12-inch calipers may be used only a few days a month on the production floor. Although they are both 12-inch calipers, their usage rates are very different, so their calibration intervals should be different. Gauge usage rate needs to be considered when determining calibration intervals. However, it is not the only factor to consider. Usage environment is the other. A caliper used in a room with high levels of vibration, grime, and seasonal temperature changes may need calibration more often than the same type caliper used daily in laboratory conditions. Consider also that some employees are rougher with gages than others and work habits of a particular department may not be as careful as another. All these things point to the fact that calibration interval is an individual thing. Just because a gauge is a certain type does not mean its calibration interval should be the same as every other gauge of that same type.

The third common strategy is for all gauges to have the same calibration interval, but be evenly spread out throughout the year. All gauges having the same calibration interval will inevitably result in some gauges not being calibrated enough while others are calibrated too often. This is the costliest and least efficient of all the interval strategies.

Typically most companies calibrate most gauges too often. Knowingly or not, they are applying the "better safe than sorry" philosophy. This may be a good idea with some situations, but it is not always a productive way to do gauge calibration. Calibration interval affects productivity in two ways: gauge availability and gauge accuracy. Unavailable gauges may cause an operator to skip a measurement or waste labor time looking for a gauge or a suitable replacement. Nonconformances caused by inaccurate gauges or skipped measurements affect productivity in a bad way by wasting time, material, labor, and money when defective ports are not usable later on.

Take the time to consider usage rate, usage environment, personnel, calibration time and complexity, and gauge storage so as to keep your gauges accurate and available. One way to do this is to sort your gauges into interval classes. Examples of interval classes may be monthly, quarterly, semiannually, and annually. Determine the calibration interval for each gauge, and then place it into the appropriate interval class. Resist the unscientific and unjustified habit of determining calibration intervals by gauge type alone.

Another way companies make it more likely to have gauge calibrations expire is to have calibrations expire on a particular day. Neither ISO 9001 nor ISO 17025 require a specific day for calibration expiration. You can have the calibration expiration date be simply a month and year. This gives you the whole month in which to calibrate the gauge. That means the calibration is less likely to come due when the department is busiest. It also means the calibration department can do the calibration at a time more convenient for everyone, and by doing so, interfere less with productivity.

If a customer or other requirement pins down a specific day, you can write into your calibration procedure that the device may be calibrated anytime during the month in which the gauge expires, or wording to that effect. It is unwise to make the rule that calibrations are due by the end of the month because that is when production is busiest and gauge recall occurring then will surely impact production activity. In any case, giving your calibration personnel and the department that uses the gauge some leeway in deciding when to calibrate a gauge can prevent such problems as a gauge being recalled when it is needed most. It will also make it easier for the calibration technician to keep current.

Unneeded calibration is another error that wastes time and resources. ISO 9001 requires calibration only on measuring devices that are used *for acceptance of product.* No other calibrations are required by the standard. So unless a toolmaker's micrometer is the exact same one being used by your final inspector, there is no need to calibrate the toolmaker's micrometer because he or she is using it for working on tooling, which is not your product. Since it is not for product acceptance, ISO 9001 does not require it to be in your calibration system. The same is true for the pressure gauge on the boiler that heats your building and for an oil gauge on a machine. Unless the machine itself is your product, it does not have to be in your calibration system. Besides, toolmakers usually do excellent jobs of keeping their own measuring devices accurate as they usually check them far more often than the calibration department would.

Some customers may require gauges to be calibrated whether or not they are used for acceptance testing. They may do this on the grounds that the calibration of tooling and equipment affects the outcome of the manufacturing process. This may be true, but the ISO standards require calibration only on gauges used for accepting *product,* not on gauges used to monitor or control *processes.* This is not to advocate that you don't calibrate them, but only that you are not required to do so under ISO 9001. If your process is sensitive to variations in such instrumentation, it may be worth calibrating them. Not calibrating gauges that signal processing problems can in some cases result in increased nonconformances, which lowers productivity.

Another calibration error that affects productivity is calibrating gauges at the wrong place or under the wrong conditions. Some companies require gauges to be brought to a metrology lab, held there long enough for temperature stabilization, and then calibrated under laboratory conditions. This is fine if the acceptance of product occurs in a laboratory under laboratory conditions. However, just because a gauge is in calibration under laboratory conditions does *not* mean it will be in calibration where it is actually used. If the product acceptance measurements take place on the shop floor, calibrating in a metrology lab under laboratory conditions makes no sense. Gauges need to be accurate *where they are used.* If a gauge is perfectly calibrated in the laboratory but off calibration on the shop floor where it is being used for product acceptance, what good is that? Inaccurate gauges can result in an increase in nonconformances and reduced productivity.

Calibration records are an important part of any calibration system. The necessary records are gauge identification, last calibration date, calibration due date, and results of the most recent calibration activity. Use some good judgment about what other information should be in the record. Recording the location of a gauge may save time in finding it, depending on how mobile the gauge is and the amount of sharing between people and departments that takes place. Time spent looking for a gauge is time not used productively. Thus lost gauges can reduce productivity.

Excessively short intervals keep your calibration people doing more work than they need to and reduces gauge availability. Both of these have an undesirable impact on productivity. Excessively long calibration intervals increase the risk of using an out-of-calibration gauge, which is equally undesirable for productivity. That is why it is wise to have a means of changing the calibration interval included in your system. A review of the calibration history of a particular gauge is necessary to wisely judge when and how to adjust its calibration interval.

The question arises as to how far back into the gauge history the calibration records need to go. Actually that depends on several factors. One thing to consider is that if you use calibration history as a basis for adjusting the calibration interval, you need to keep records going back far enough to give the necessary history. For example, your system may allow for a 30% increase in calibration interval after five calibration cycles where no adjustment was needed. If the calibration cycle for the gauge is one year, then you need five years of history in your calibration record. The amount of history needed is therefore a record retention period that is determined by your system. For very old gauges that have 30 years of calibration history, it might make more sense to record the procurement date of the gauge and then keep only the past 7 or 10 years of calibration records. In some companies, calibration record keeping is often a matter of keeping only the most recent calibration record, but that prevents using calibration history for changing the calibration interval.

SAMPLING

Documentation issues, calibration practices, and SPC are not the only quality activities that affect productivity. Sampling methods themselves can have an impact on productivity by lowering time efficiency. It takes time to do measurements, so the best use of that time must be made. One way to make the best use of time spent of measuring is to use the most efficient sampling plan.

Sampling affects productivity by determining the amount of time spent measuring product rather than manufacturing it. Some sampling plans are more efficient than others. By *efficiency of sampling plan* is meant the number of samples needed to draw the conclusion with the desired level of confidence. A sampling plan that requires 50 samples to be 95% sure that the lot of a given size is good is less efficient than a sampling plan that gives you 95% confidence but requires only 32 samples for the same lot size.

Generally speaking, sampling by variables requires smaller sampling sizes than sampling by attributes. Plans of the $C=0$ type require even fewer samples. However, you must pay attention to the sampling error. There are two types of sampling errors that can be made: The first is to reject something that should have been accepted. The other is to accept something that should have been rejected. Both of these increase as the ratio of sample size to lot size gets smaller.

Rejecting something that should have been accepted is known as producer's risk. It is expressed as a percent of the times this error is expected to occur and is designated as alpha (α). Accepting something that should have been rejected is the consumers risk. This, too, is expressed as a percent of the number of times it is expected to happen with a given sample size. It is designated as beta (β). The sum of these two error rates is known as sampling error and typically should be less than 10%, which gives you 90% confidence in the conclusion drawn from your sample. Average quality level (AQL) type sampling plans as in ISO Z1.4 typically have low alphas, so they are often used by manufacturers when inspecting their own product. ANSI or ISO Z1.4 has superseded MIL-STD-105 E and provides a variety of sample plans of lot-by-lot inspection that are specific to processes done in a series of lots or batches. Rejectable quality level (RQL) plans, also known as LTPD sampling plans, typically have low betas and so are recommended more for incoming inspection where you are inspecting what someone else has produced. These are best applied to isolated or infrequent lots.

C = 0 sampling plans as described in MIL-STD-1916 are often used. MIL-STD-1916 has superseded MIL-STD-105 E and provides a variety of sample plans of lot-by-lot inspection, as well as continuous inspection plans that are specific to continuously running process that are not done in lots or batches. Like MIL-STD-105 or ANSI Z1.4, this standard also contains sampling plans that can be used as either AQL or RQL sampling. They typically have a total sampling error averaging around 90% and require the smallest sample sizes. Since C = 0 type sampling plans use smaller sample sizes, it takes less time to measure the complete sample quantity. There are also continuous sampling plans, which actually apply very often in modern manufacturing and are among the most efficient. These are recommended for processes that have a continuous flow of product that is not divided into lots or batches.

When sampling by attributes, double and multiple sampling plans like those described in the now-canceled MIL-STD-105E or in ANSI Z1.4 can result in making the accept or reject decision while inspecting considerably fewer samples than single sampling.

Double sampling in attribute sampling schemes begins by pulling a small sample and accepting the lot if the number of defects in the sample is below a certain number, known as the accept value, or rejecting it if the number of defects is above a different number known as the reject number. Next, if the number of defects is between the accept number and the reject number, an additional sample is pulled and inspected to make a decision. For example, your lot size is 1,000 and your single-sampling plan calls for 80 samples, so you accept on 1 defect and reject on 2. If you use a double-sampling plan instead, then you can use smaller samples. For a lot size of 1,000, you may have to take a sample of only 50 pieces. You would then accept on zero defectives and reject on 3.

If you had only 1 or 2 defectives, then would you have to take a second sample of 50, in which case you would accept on 1 and reject on 2. This usually occurs on a minority of the lots sampled. Most often the 50 piece sample is sufficient. This is a difference of at least 30 samples that need to be measured, since if you find no defectives the 50 pieces would be sufficient, as opposed to always having to pull a sample of 80. Imagine the time savings of inspecting up to 30 fewer sample pieces most of the time. Multiple sampling plans work in a similar fashion but have two or more pairs of accept and reject numbers with additional sampling required. Multiple sampling plans are sometimes called sequential sampling plans. They are rarely used anymore and even more rarely published.

But this is by no means the only way to make sampling more efficient. Switching rules can also reduce inspection time by making sampling more efficient. Using the switching rules in either ANSI Z1.4 or MIL-STD-1916 results in fewer samples and less time inspecting, hence, productivity improvement results.

Switching rules, whether applied to attribute- or variable-type sampling schemes, result in less time spent measuring. Switching rules allow for a reduction in sample size based on the quality history of the item being inspected. For example, if you are taking a 32-piece sample and have not found a defect in the past 10 lots, then you can switch to a 20-piece sample and continue to inspect only 20 samples per lot. You remain at a sample size of 20 until a defect is found, and then you go back to inspecting 32 samples again. Switching rules also apply to a worsening of quality history. If you are inspecting 32-piece samples and find 5 rejected lots, then you must increase the sample size to 50 and remain there until 10 lots are found with no defect. Then you can revert back to the 32-piece samples. Switching is more efficient than constant single sampling and makes better use of inspection labor hours. However, it is applicable only when you are inspecting a continuing series of lots from the same source. Switching rules are not applicable to isolated or infrequent lots and should not be used for them. They are also not for situations where lots from different sources are combined.

Besides sample quantity, inspection time can also be reduced by the choice of gauging. Simple and quick to use, attribute gauges can be faster than measuring with variable-type gauging. Go/no-go gauges can be custom made and quickly identify a pass or fail condition. For space dimensions, a go/no-go gauge can be a solid negative of the space and act as a mating part. By making the gauge a mating part with multiple contact surfaces, several space dimensions can be checked at once. Likewise, for material dimensions, make the gauge a negative of the material shape having space dimensions that check the fit of the part. By using custom fixtures, operators can check multiple dimensions with a single motion using a single gauge. That is a real time saver.

Sometimes inspection takes less time if it is simply more convenient to do. Designing and making fixtures that hold the part a certain way may reduce inspection time by simply being more convenient. Anytime you can reduce inspection time without compromising quality, it can improve productivity. If the operator is doing the inspection, reducing inspection time makes labor efficiency go up due to less time being spent checking parts and more time producing them. If the inspection is done by an inspector, decreasing inspection time per sample will improve the inspector's throughput, which increases material flow, resulting in better productivity.

LACK OF FOLLOW-UP ON CORRECTIVE AND PREVENTIVE ACTIONS

The *act* portion of the PDCA cycle is critical to productivity improvement. If it is not done properly, productivity will be reduced. Therefore, timely and correctly performed follow-up to any corrective or preventive action is advisable to maximize the improvement to productivity. Follow-up on corrective and preventive actions consists of three things: First, verify the completeness of the action itself. Second, verify the proper and complete implementation of the action. Third, verify its effectiveness.

A complete corrective action must have both a long-term and a short-term solution. The short-term action has two purposes: The first is to shut off the symptoms so they never reach the customer, neither the internal customer nor the external one. The second purpose is to contain the problem so that it does not spread. In other words, limit as much as possible the amount of product or work in progress affected by the nonconformance.

The long-term action is to eliminate the root cause or causes. An important step in eliminating the root cause is to verify that it actually is the correct cause and not just someone's pet peeve or favorite theory. If someone has a favorite theory about which they are quite certain, conduct an experiment or do an investigation of some kind that will either prove or disprove it. There is after all a possibility that they may be right. Always take the time to verify the actual cause or causes; otherwise, you may injure productivity by implementing a change that did not have to be made and may in fact reduce productivity. As part of the permanent corrective action, do not forget to consider and include a remedy for whatever it is that allowed the root cause to exist in the first place. This is known by various names—systemic action, managerial action, operational action, and others.

After verifying the completeness of the action, the next part of following up is to verify that the action was properly and completely implemented. Do not assume that because someone was told to implement such-and-such an action that it was done the way the action developer intended. Even with the best of intentions, miscommunications, differences in perception and priorities, and differences in work experience can result in the action not being implemented as originally conceived. Verification is an important follow-up that should not be skipped.

The verification is not merely a check to see if it has been implemented, Rather it is a verification that the implementation is done as intended by those who developed the corrective or preventive actions. The verification should be done by the individual or team that developed the action in the first place, since they would know best what was intended.

When verifying implementation, do not forget to check that any operator training that the action made necessary was satisfactorily completed. Be sure the applicable supervisor is well aware of any differences in the process or equipment that is being used. Check that any new or revised tooling is properly identified, and that the effective date is accurate. Change any applicable paperwork such as operations sheets, shop routing documents, and forms to be filled out. If the process is carried out on more than one shift or by more than one team, be sure to adequately verify each of them.

Verification of implementation is only the second part of the necessary follow-up. You must also perform the third part of follow-up, which is to verify the effectiveness of the corrective or preventive action. The effectiveness is ascertained by measuring or counting some output before the corrective action that can be compared to what it is after the corrective action is implemented. Thus all corrective and preventive actions must have a measureable or countable output that can be used to verify their effectiveness.

This output can be as simple as counting the number of defects before and after implementation and comparing the two counts. Many companies graph outputs and post the graphs where they can be seen by everyone. This not only shows the effectiveness of the action, but it may be a source of motivational pride when everyone can see how well the action is working.

Sometimes the effects of an action are not dramatic enough to be obvious or otherwise require some statistical analysis to be seen. Comparing process capability, stability, or both from before and after the implementation may best show whether or not the action is effective. More subtle improvements or measurements that are subject to a lot of statistical noise, variation, or multiple influences may require more sophisticated statistical analysis, such as tests of significant difference, correlation and regression analysis, or other statistical methods.

In this case it is important to do the right statistical analysis so as to prevent an incorrect conclusion from being drawn. Table 5.1 shows what statistical analysis to use under which conditions.

TABLE 5.1

Types of Statistical Analysis Useful to Productivity Improvement

Analysis	Usefulness
Analysis of Variance (ANOVA)	Compares results of different groups like fixture to fixture, shift to shift, etc., to determine where the greatest variation is. Wherever the greatest variation is, the need for control will be greatest to prevent defects.
Correlation	Disproves cause and effect if the correlation is low. High correlation implies cause and effect, but experimentation is needed to prove it.
Control Chart	Tells if a process is stable and when the process needs adjustment.
Design of Experiments (DOE)	Ranks process parameters by their degree of effect and shows their degrees of interaction. Also tells optimal settings for the tested parameters.
Process Capability	Tells if a process can meet the specification. It can also be used to predict defect rate.
Sign Test	A test to determine if there is significant difference in data before and after corrective action. It is applicable to any distribution type.
t-Test	A test of significant difference that is applicable only to normally distributed data. Most *t*-test tables allow for varying degrees of sensitivity that the user can select.

Any time a preventive or corrective action is not completely followed-up on, the potential exists of failing in correcting or preventing the problem. Since it is the problem, that is, the failure mode, which affects productivity, failure to adequately correct or prevent it is detrimental to productivity. Productivity improvement requires containing the failure mode and preventing it from ever occurring again. Without proper follow-up on corrective and preventive actions, you cannot know if the action taken did indeed improve productivity.

6

Productive Manufacturing

WORK IN PROCESS

Work in process, commonly known as WIP, can affect productivity in different ways. One way is by the inefficient transportation and availability of material, parts, and subassemblies. *Where* material and parts go is important, but *when* they get there is equally important. How and in what quantities are also worth looking at.

Efficiency of transportation is to some extent a matter of routing. Moving materials and parts from one department to another is inevitable, and very often the location of these destinations causes much back-and-forth movement. Changing the plant layout is usually not an option, so the quantity and timing of the movements becomes more important. In an ideal situation, a lot or batch should be one day's work. This is fine if a single job order is for a large enough quantity to be several days' worth of work. But this is not always the case. If a single job is about one day's work, it may be most efficient to move the material all at once. If job orders are small, they may be combined, although this may risk loss of traceability if that is a requirement. The point here is that the movement quantity should not be arbitrary, but chosen for economy of movement and maximization of material flow.

The same principle applies to moving material through a single work cell. Within a cell there are actually two options: lot or batch movement, or single-piece continuous flow. If movement in lots or batches is being done, the lot quantity is important. Again the ideal situation would be one day's production, one job order, or several orders at once, depending on the situation.

If one-piece continuous flow is being done, the flow should be timed so that work does not pile up at the slowest operation. To prevent idle waiting times, operations can sometimes be combined. Bottlenecks can in some cases be relieved by duplicate workstations. Sometimes an operation is slow enough that one operator can do more than one workstation. Operation timing is the key to productivity with one-piece continuous flow. In any case, operator input is at least as valuable as an observer with a stopwatch, so discuss proposed changes with operators before implementing a change.

Besides efficiency of movement, movement quantities, and movement timings, you must also look at locations and identification of materials, parts, and subassemblies. Misidentification causes movement to the wrong place or results in no movement at all. The location must also be clearly identifiable and accessible and is just as important as the material and parts identification. Many companies are very good at identifying locations but not as good at keeping them accessible. Stopping the flow of material to gain access to a location is not a productive use of anyone's time.

If the operators have to get their own parts, all the necessary parts should be not only well identified but also easily reachable without having to walk around to get them. If parts are brought to the operators, they should be dropped off at the place most convenient *for the operator*, not for the materials handler.

Of course, all bins, boxes, or other containers for materials, parts, and subassemblies should be not only easily reachable but clearly and completely labeled so that mistakes are less likely to occur.

Control over the manufacturing processes is yet another way manufacturing affects productivity.

Different processes often require different amounts and types of control. New processes may need to be tightly controlled, at least initially. However, process control methods can and should be periodically reviewed and adjusted to get maximum benefit for the controls being used. Overcontrol is a waste of resources and is counterproductive. A good quality history can justify smaller inspection and testing sample sizes. With less time being spent on inspection and testing, the assembly line may operate more quickly and efficiently. Likewise, insufficient control increases wasted time and materials by increasing the proportion of defective items.

Manufacturing engineers know that the order in which operations are performed is important in some designs and processes but not in others. If operators are locked into a particular sequence when the sequence is totally irrelevant, productivity can decrease. The loss of productivity can occur in two ways. First, if parts are not available, the process must be stopped while waiting for the parts. If the sequence of operations is truly not relevant, the operator could do some other operation while waiting to receive the parts. Of course this would not apply in situations where the operations must be performed in a certain order.

The other way that sequence of operations may decrease productivity is if the sequence of operations does not match the layout of the work cell. This causes wasted movement of materials and people. The remedy for this is to either change the work cell layout or, if practical, allow the operators to do the operations in the order most convenient to them.

EFFECTIVE VERSUS INEFFECTIVE STATISTICAL PROCESS CONTROL

Statistical process control, also known as SPC, can be effective at maintaining productivity and even improving it. However, this is true only when the SPC is properly applied. It must be properly implemented on the right characteristics in a timely manner. SPC can reduce variation and in doing so prevent defective product from being manufactured. The reduced variation and lower defect rate contributes to increased productivity. SPC also can provide helpful information for developing effective corrective actions that reduce the defect rate and so contribute to a higher level of productivity. Many companies use it well and truly benefit from doing it. Benefiting from SPC requires that it be properly implemented. It must be applied wisely and be properly performed.

However, there is widespread misunderstanding about what SPC is supposed to accomplish, along with where and when to use it. Due to ignorance and incomplete understanding of how to begin SPC, it is often incorrectly implemented and so fails to meet expectations. It ends up being a burden on productivity rather than a help. SPC can be good for productivity only when it is properly applied and performed.

There are several reasons why SPC will not benefit productivity. Some of the most common are

- SPC is not being applied on the right characteristics or the right manufacturing process.
- Nonrandom variation was not eliminated, or process stability was not achieved *before* starting SPC.

- The time lag is too great between the time of manufacture and the time the measurements are made, or from the time of measurement to the time of plotting the chart.
- The wrong kind of SPC chart is being used.
- The charts are not being examined each time a measurement is plotted.
- The person doing the SPC is not the person controlling the process.
- There is no response to the SPC chart when it calls for process adjustment.

Determining When and Where to Do SPC

Many companies are going through the motions of SPC but receiving no benefit from it. In such companies, the operators may consider it a waste of time. Sometimes it provides fertile ground for malpractice. Companies that are only going through the motions of SPC usually collect the SPC data, file it, and then ignore it, or show it to auditors and customer representatives.

Applying SPC correctly first requires the proper determination of when and where to do SPC. Some companies foolishly try to do it everywhere. It does not take long for those companies to realize that this is a mistake. In this situation management soon sees no benefit from most of the SPC. Consequently SPC is seen as a waste of effort and is soon abandoned. Other companies try to use SPC to control product dimensions selected as more important than other dimensions and often identified on a print. They are in effect using SPC as an inspection method rather than a process control method. Granted, there are situations where dimensional measurements made on a part will indicate that the process needs to be adjusted. But these dimensions must be chosen for what they tell you about the process and not on the basis of dimensional criticality or safety. This misapplication of SPC does little to improve productivity. SPC is a *process* control method, not a *part* control method.

There are some processes for which SPC is not a good choice as a control method due to the lack of operator-controllable variables. For the operator to control a process, three criteria must be met: The operator must know what the part characteristic is supposed to be, must have a way to compare the actual characteristic to a standard or specification, and must have a means of adjusting the process to reconcile any difference between the actual characteristic and what it is supposed to be. Unless the operator has all three, that person does not have control over the process, and having him or her do SPC will not enhance productivity.

An example of this is stamping metal parts with a die in a press. If a die-stamped part comes out of the press with a dimension that is out of tolerance, what can the operator adjust? Dimensionally out-of-tolerance stamped parts usually indicate a problem with the die, which must be removed and worked on by a skilled craftsman, such as a tool and die maker, toolmaker, or skilled machinist. Neither of these is something that can be adjusted by an operator in response to an out-of-control indication on an SPC chart.

SPC dimensions can be chosen in several ways. Be cautious here. Select such dimensions for SPC only if the operator can make an adjustment to bring them back into control. If a dimension is not operator controllable, there will be no benefit to doing SPC.

To correctly determine on which characteristics to do SPC, first check the process failure mode effects analysis (PFMEA). If a failure mode has as its cause something about the process that can be adjusted by the operator during production, and the risk priority number (RPN) is at or above the preventive action threshold, then it is a good candidate for SPC. If the RPN is above the action threshold but the cause is not something adjustable by the operator, then some control method other than SPC should be put into place.

Control Plans and PFMEAs

Control plans, when developed and implemented properly, are a useful and effective tool to maintain the productivity of a process by keeping that process under control so that time, material, and labor are not wasted. Unfortunately, some companies produce control plans just to satisfy a customer and do not actually use them for productivity. To be effective as a productivity tool, a control plan must be kept current as a living document, properly developed, and actually used by production. It is useless as a productivity-enhancing tool if it is filed away in a drawer or on a computer.

Control plans are supposed to be made from the PFMEA and the first-article inspection results, but before SPC is started on production. Actually, it is the control plan itself that tells the operator on what characteristics to do SPC. The characteristics that are to be controlled are listed on the control plan. They are determined by the RPN being at or above the threshold RPN number on the PFMEA and by the characteristics on the first-article report that are right on the tolerance limit or have low Cpk. The control plan lists these characteristics and tells what control method is going to be used by the operator to keep the process in control, so that these characteristics do not go out of specification. The control plan is an *operator's* document. It is meant to be used by the operator to control the process. Therefore it must be written in a way that the operator can understand it. The control plan tells the operator how to do the measurement, on how many samples, and how often. Most importantly, the control plan tells the operator what to do when the process goes out of control—that is, when a measurement or visual characteristic is not within specification or when an SPC chart indicates a process adjustment is needed. What to do when a characteristic is out of tolerance, or out of control, is shown on the control plan. It should be posted at the workstation and easily visible to the operator. The operator must be trained to use the control plan. It is useless if it is filed away somewhere. Figure 6.1 shows an example of a control plan.

As manufacturing continues, quality problems not foreseen during the development of the PFMEA and first-article inspection may occur. Corrective and preventive actions are developed to deal with these problems. This is when the PFMEA and the control plan get updated. Each time a new failure mode occurs and a corrective or preventive action is developed, update the PFMEA to show the new defect mode and its corrective or preventive action. Then calculate its RPN. If the RPN is at or above the control threshold, add to the control plan whatever is necessary to identify the new defect mode and what needs to be done to prevent the process from producing that defect mode again.

When defect modes already on the PFMEA are seen in production or found by the customer, the probability and occurrence estimates may need to be increased if they occurred more frequently than expected or if they are noticed by the customer more easily than anticipated. Problems that are not already on the PFMEA need to be added and have their RPN calculated. Add corrective or preventive actions for them if their RPN is high enough to exceed your action threshold. If problems occur that are already on the PFMEA and preventive or corrective actions have already been taken, this may indicate that the preventive or corrective action is not effective and needs to be revised. In either case, every time you revise the PFMEA, review it to determine if any revision of preventive actions is also needed. Review and revise the control plan as well.

Customer complaints are another source of information that must trigger a revision of the PFMEA, the control plan, or both. Whenever the preventive action on the PFMEA is changed, the action on the control plan ought to be reviewed because it may also need to be changed. This is why PFMEAs and control plans are considered living documents. They are constantly being updated as needed to keep the process in control,

		PROCESS CONTROL PLAN				Date:	6/10/2010		Page 1 of 1
PART NO:		1234567-9				WORK CELL	#7		
PART NAME:		Faceplate	PROCESS STEP:	Drill mounting holes		NOTES:			Appearance is critical to customer
Characteristic	Specification	Control method	Sampling	Gauge type	Gauge R&R	Action plan when defect occurs			
Mounting hole size	.270 ± .010	1st-pc inspection	1 pc	Optical comparator	1.3% of tolerance	Stop process, check setup, and verify proper tooling			
Hole locations	Position tolerance .005"	SPC	3 pcs every 2 hours	Optical comparator	1.3% of tolerance	Stop process, clean fixture, verify indexing			
Appearance	No scratches or dents	Visual inspection	100%	N/A	N/A	Scrap piece and replace			

FIGURE 6.1
Control plan example.

so as to identify defects as early as possible in the process or, better yet, prevent them from occurring in the first place. Well-developed and properly implemented PFMEAs and control plans result in fewer defects being manufactured. Consequently, less time, material, energy, and other resources are wasted. This reduction in wastes increases productivity.

If dimensional measurement must be used to indicate how well the process is doing, then another way to determine which characteristics should be controlled by SPC is by doing a first-article inspection and process capability study. To do this, check the first-article inspection data to see if any dimensions are right on a specification limit. Doing SPC on only those dimensions that are right on the specification limit will prevent defects. Alternatively, you can determine on which dimensions or process characteristics to do SPC by the process capability for the dimension. SPC is considered necessary on processes having a Cpk less than 1, or in some industries, 1.33. Still another way to identify on which dimensions to do SPC is to select the dimensions that affect the ability of the part to be assembled. These will typically be geometric dimensions like true position of a hole, flatness, straightness, profile of a surface, and so forth.

Besides doing SPC on the wrong characteristics, another common mistake that causes SPC to be ineffective is to start the SPC without first eliminating the nonrandom causes of variation. Such causes are also known as assignable or special causes. SPC assumes a normal data distribution curve (bell curve), and your measured data will not exhibit this if assignable causes of variation are at work in your process. Examine the shape of the histogram made from a sample of about 50 pieces to get a good idea of the shape of the distribution curve that applies to your data. Consult with publications on SPC or other statistical sources for making and interpreting histograms, if necessary.

Starting SPC

It is true that some processes and part characteristics have frequency distributions that are not intrinsically normal, but these are not common. A vast majority of measureable physical characteristics are indeed normally distributed when special causes of variation are not present. Analysis of variance, gauge repeatability and reproducibility (GRR), designed experiments (DOE), six sigma projects, and others are tools to help eliminate the assignable causes that affect a process and prevent it from being normally distributed. A second reason for eliminating assignable or special causes of variation is that their elimination is necessary for the process to be stable.

Flatness and a few other characteristics typically do not exhibit a normal distribution curve. Also, a distribution curve may not be normal if it is not possible to remove an assignable cause for reasons of technology or design. It these cases, remove whatever assignable causes you can. Then you can normalize the data by using X-bar and R charts. With X-bar charts, the data plotted are actually the average of the raw data for the subgroup, so the central limit theorem will cause the data to be normally distributed. Subgroup sizes of 2 to 5 are most common, and rarely does subgroup size need to be greater than 10. The further from the classical bell shape your raw data are, the larger the subgroup you will need.

After these causes have been eliminated, you can begin SPC charting. Begin by selecting the right kind of SPC control chart. After the choice of SPC chart has been made and enough subgroups of data have been collected, it is time to determine stability by making a trial control chart. Be sure to have enough data collected for valid control limits; typically this would be 10 to 20 subgroups. For best results and to ensure validity, 20 subgroups are recommended.

This trial chart has temporary, trial limits by which stability is determined. If the chart shows that the process is not stable, then assignable causes are corrected by process modifications and a new set of 20 subgroups is used on a new chart to determine stability. When stability is established, the temporary control limits become fixed limits and may be used until the process is modified or an out-of-control situation occurs. Formulas for calculating control limits for variables-type charts where data are measured rather than counted are given in Table 6.1. The factors needed to apply the formulas for the control limit calculations are given in Table 6.2. Control limits are calculated from these factor tables. Use A_2 for X-bar charts, E_2 for X charts, D_3 and D_4 for R charts, and B_3 and B_4 for S charts. S charts are recommended if the number of samples in a subgroup is greater than 10. For attribute-type control charts, where the data plotted are counted rather than measured, the formulas used to calculate the control limits are given in Table 6.3.

A stable process is important for proper calculation of process capability and proper calculation of control limits. If the process is not stable, your process capability calculation (e.g., Cpk) will be only a snapshot of a constantly changing value and therefore will not be necessarily representative of the process at any point in time.

TABLE 6.1

Formulas for Calculating SPC Chart Limits

Statistic	Formula	Description
MX-bar	$\sum_{i=1}^{N}(X_i)/N$	Moving average is calculated from N number of sequential values advancing by one value for each MX-bar. Example: if $N = 5$, then first MX-bar is calculated from values 1 through 5, the next MX-bar is calculated from values 2 through 6, the next with values 3 through 7, and so on.
MR	$\sum_{i=1}^{N}(R_i - R_{i-N})/N$	Moving range is calculated as the absolute value of the differences between maximum and minimum within groups of values of group size N, in an advancing manner similar to MX-bar.
MR-bar	$\sum_{i=1}^{N}(MR_i)/N$	Average of the MRs.
R	$X_{max} - X_{min}$	Range is the difference between the highest value and the lowest value within a subgroup.
R-bar	$\sum_{i=1}^{N}(R_i)/N$	Average of the ranges.
X-bar	$\sum_{i=1}^{N}(X_i)/N$	Average of the measured values not calculated in advancing groups, but simply the average of all the X values.
S-bar	$\sum_{i=1}^{N}(S_i)/N$	Average of the standard deviations of the values within the subgroups.
UCLx	X-bar + RA$_2$	Upper control limit of X or X-bar is X-bar + (R multiplied by A$_2$).
LCLx	X-bar – RA$_2$	Lower control Limit of X or X-bar is X-bar – (R multiplied by A$_2$).
UCLr	R-barD$_4$	Upper control limit of ranges is R-bar multiplied by D$_4$.
LCLr	zero or R-barD$_3$	Lower control limit of ranges is zero at $N \leq 8$. otherwise R-bar is multiplied by D$_3$.
UCLs	S-barB$_1$	S-bar multiplied by B$_4$.
LCLs	S-barB$_2$	S-bar multiplied by B$_3$.

Control charts of unstable process cannot correctly identify and indicate stability. Control charts made on unstable processes will show that the process is out of control or, if made over too short a time, will falsely indicate that it is in control. To be in a state of statistical control is to be stable. Another reason for having stability before you begin SPC in production is

TABLE 6.2

Factors for Control Chart Limits

N	A_2	E_2†	D_3	D_4	B_3	B_4
2	1.88	2.66	0	3.27	*	*
3	1.02	1.77	0	2.57	*	*
4	0.73	1.46	0	2.28	*	*
5	0.58	1.29	0	2.11	*	*
6	0.48	1.18	0	2.00	*	*
7	0.42	1.11	0.08	1.92	*	*
8	0.37	1.05	0.14	1.86	0.185	1.815
9	0.34	1.01	0.18	1.82	0.239	1.761
10	0.31	0.98	0.22	1.78	0.284	1.716
11	0.29	*	*	*	0.321	1.679
12	0.27	*	*	*	0.354	1.646
13	0.25	*	*	*	0.382	1.618
14	0.24	*	*	*	0.406	1.594
15	0.22	*	*	*	0.428	1.572

* Chart type is not recommended for subgroups of this size.
† For X – MR charts the N is the moving range group size. Although the X subgroup size is always 1, the value of E_2 changes according to the MR group size.

TABLE 6.3

Calculating Control Limits for Attribute Control Charts

Chart Type		Control Limit Calculation
p chart	$p \pm 3 \sqrt{p(1-\bar{p})/n}$	where p is the average percent defective in a sample and n is the sample size
np chart	$np \pm 3 \sqrt{np(1-p)}$	where np is the average number defective units in a sample and n is the sample size
u chart	$u \pm 3 \sqrt{u/n}$	where u is the average number of defects found regardless of sample size and there may be more than one defect per sample

that if you calculate control limits for an unstable process, the control limits will be so wide as to be ineffective, or they will not fit your data for long enough a time to do you any good. Only when the process truly shows stability do you finalize the control limits and begin charting production. Processes in control will remain within the control limits and will not exhibit any trends, shifts in the average, or other nonrandom patterns.

When the control chart shows the characteristic to be out of control, it is telling you to adjust the process. If the process is not adjusted, it may stay out of control or even get worse. The loss of control increases the probability of manufacturing the parts out of tolerance, which causes an increase in scrap and a corresponding decrease in productivity. Many companies provide operators with a predetermined set of corrective and preventive actions to restore control when a chart indicates that the process needs to be adjusted. Such action plans are often part of a process control plan or may simply be a list of problems and appropriate actions that is posted at the workstation.

If there is a time lapse between measuring the characteristic and plotting the measurement on the chart, then the SPC will not be effective. The pieces manufactured during that time lapse may be of unknown status because the operator won't know if the process is out of control until the measurement is plotted on the chart. If the chart shows a trend heading toward a specification limit, the trend can continue during the time lapse. This may result in defective parts already being manufactured out of tolerance by the time the operator finds out about the trend. Because all the parts manufactured are now suspect, they have to be sorted, and this significantly decreases productivity. Both the time used up in sorting and the materials wasted making bad parts are not recoverable, nor is the loss of productivity. Therefore, to do SPC effectively, the SPC samples must be measured and plotted as soon as possible each time the samples are taken. The control chart must then be examined right away. Any time lag in measuring, plotting, or responding to the control chart causes parts to be manufactured at risk.

Some operators feel they do not have time to plot charts during production, or their supervisors have told them to do it at the end of the shift. This defeats SPC. It prevents SPC from working and renders it a useless waste of time for reasons described in the preceding paragraph. You cannot do SPC effectively that way. If SPC is going to aid productivity, the measurements, the plotting of the measurements, and the examination and response to the control chart must all be immediate. This is so that the operators can respond to the chart in time to prevent out-of-control manufacturing. Keeping the manufacturing process in control is how SPC aids productivity. Processes in a state of statistical control have less variation, and by keeping the process in control, defects are prevented.

Another error that causes SPC to be ineffective is using the wrong kind of SPC chart. Certain kinds of SPC charts are not effective unless the data for the characteristic being controlled is normally distributed. To use these charts, it is necessary to eliminate assignable causes of variation. One kind of chart for which a normal bell curve distribution is absolutely necessary is the chart for individuals. It is often paired with a moving range chart. These are called IX-MR charts or X-MR charts, and they can give you false indications of out-of-control situations if they are used with data that are not normally distributed. These charts are easily identifiable by the fact that they have a subgroup size (sample size) of just 1. Another kind of chart, X-bar and R charts (also known as average and range charts), can be used with either normal or nonnormal distributions. The further away the data are from a normal distribution, the larger the sample size must be for an effective X-bar chart. For many instances a sample between 2 and 5 is sufficient. But distribution with considerable skew or other nonnormal condition may require a sample size of 7 or even 10 pieces. X-bar and R charts lose effectiveness as you approach samples sizes of 15 or more. As a rule of thumb, if you need a sample size larger than 10, use X-bar and S charts instead. These are also called average and standard deviation charts.

Attribute control charts may reduce the amount of time spent on SPC and so give the operator more time to make parts. In attribute SPC you do not plot the measure of a characteristic on a sample of parts. Instead, you count the number of defective parts in the sample and plot that count on a chart. *P* charts plot the percent of the sample pieces that are defective. Charts called *np* charts plot the actual number of defective parts in the sample. *U* charts plot the number of defects found without regard to the number of defective parts. If one part has three defects and another has only one, the *u* chart would indicate that as four defects even though only two parts were inspected. Attribute charts are interpreted and acted on in the same way as X-bar and R charts or X-MR charts. See Table 6.4 on how to decide what kind of SPC chart you should use under which circumstances.

If the charts are not examined for out-of-control conditions every time a point is plotted, the SPC is not being performed effectively. Of course the person plotting the chart knows immediately if the current data point is out of control, but that is not the only criterion to look for. SPC charts actually signal the need to investigate the process in several other ways, all of which need to be checked whenever the chart is updated.

One thing to look for is trends. This is a group of seven or more consecutive data points all going up or all going down. If a trend is left to continue, defective parts can result. Even if the last point on the trend is not near a specification limit, the trend still indicates a loss of normality and therefore a change in the process. Find out the cause of the trend and take

TABLE 6.4

Determining Which SPC Chart Type to Use

Variables Charts (data are measured, not counted)	
Subgroup Sample Size	**Preferred Chart Type**
1	X and MR
2 through 6	X-bar and R
7 through 10	X-bar and R or X-bar and S
11 through 15	X-bar and S

Attribute Charts (data are counted, not measured)	
Condition	**Preferred Chart Type**
Counting the number of defective units of product with same defect	*p* chart
Counting the number of defective units of product with different defects	*np* chart
Counting the number of defects without regard to number of units	*u* chart

appropriate action to prevent an out-of-tolerance condition and renormalize the process. Some people think that if a trend is approaching nominal, they should leave it alone. This is true if you also identify why the trend is happening. You can allow the process to go toward nominal on its own, as this does improve quality, but something had to happen to cause the trend. If you know what it was, you can use that knowledge to learn how to center the process to nominal whenever you want, rather than hoping it will happen on its own. You can also stop the trend if it goes past nominal. This will be harder if you did not first investigate why the trend was happening.

Another thing to look for is a sudden shift in the average. Seven or more data points all above or all below the average line indicates a shift. Again, find out why it occurred and correct it. If the shift is toward nominal, let the shift stay, but find out why it happened so you can create a shift intentionally when you need to center the process, or to prevent out-of-control situations from happening in the future it if it shifts too far out.

Repeating patterns of any kind always have a periodically recurring assignable cause. Periodic repeating patterns that suddenly start on their own, or even stop on their own, indicate something is at work influencing your process. Investigate the cause of the pattern and correct it. See Figure 6.2 for examples of signals on a control chart that a process adjustment is needed.

There are some types of additional criteria used when examining SPC charts that indicate the data are no longer normally distributed. But these

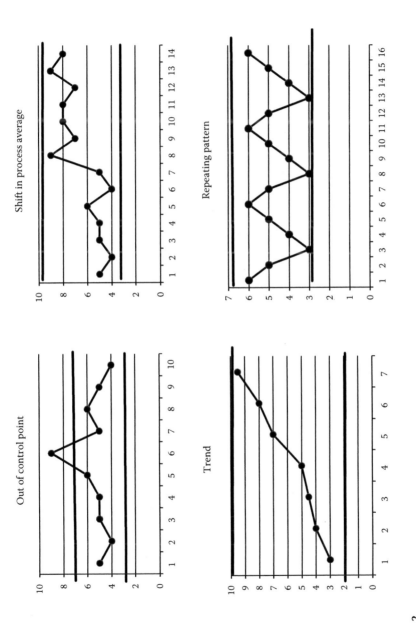

FIGURE 6.2
Examples of control chart signals indicating adjustment is needed.

criteria are not applicable to all charts or all characteristics. If your data are truly normally distributed and you want to keep it that way, then use these normality-loss-indicating criteria. Consult an SPC publication for further details on chart normality-loss indications.

Selecting SPC Personnel

The person who can control the process is the one who needs to see the charts every time they are updated. SPC will not work if one person does the measuring and potting while another person adjusts the process. These must be the same person because the person controlling the process needs to respond to the chart immediately.

An incorrect and useless way to do SPC is for the operator to pull samples, then hand them off to someone else to measure and plot the chart, and if they see an out-of-control condition, they tell either the operator or someone else. This is ineffective SPC because of the time lapses between pulling the samples and measuring, between measuring and plotting, and between plotting and responding to the charts. Due to this time lapse the parts manufactured are suspect and need to be dispositioned. What if the person doing the measuring is busy with a task that has a higher priority? Or it's break time or lunch? What if he or she is out sick that day? Even if the samples are measured and plotted immediately and an out-of-control condition is noticed, what if they can't find the operator, or if the operator is busy doing something else. It is always better and much more effective to have one and the same person pull the samples, measure them, plot the data, and respond to the charts. Not only is this the correct way to do SPC, it is the only way SPC can be consistently effective.

When SPC Calls for Action

Sometimes supervisors or people who know the process choose not to respond to the charts because the data show that all the parts are still in tolerance, so they see no reason to adjust the process. Operators and supervisors who are concerned only with making the rate or quota do not want to use time adjusting the process to fix an out-of-control condition when all the parts are good and are not in danger of going out of tolerance. People who think this way do not understand that SPC is not about keeping product within the specification. SPC is about controlling the *process,* not the *product.* Controlling the process reduces variation, reducing variation prevents nonconforming parts from being made, and preventing nonconforming parts will make it easier for the employee to make their quota or rate.

Others who think that the process can be left alone as long as everything is within the specification may not fully believe in SPC or truly understand it. Some consider it "statistical mumbo jumbo." Those who do not believe in or understand SPC will also usually not believe that it is more economical to prevent defects than to rework or replace them. Both ideas are wrong. SPC has been around since the 1920s and has been proven time and again to be an effective productivity-enhancing tool, but only when applied to the right processes, properly initiated, and properly performed.

Besides examining the control charts for out-of-control conditions, the process capability should be periodically checked. The better the capability is, the more productive the process will be. This is because increased capability means fewer defects. However, process capability calculations can be misleading if the process is unstable, if the wrong capability index is used, or if the capability calculation is incorrect for the type of distribution.

When selecting a process capability measure, you also must determine how the specification requirement is stated and whether the process is just starting up or is ongoing. The process capability calculations for ongoing processes are measured as Cp, Cpk, or Cr. These typically use the standard deviation as estimated by the R-bar/d_2 method if they are normally distributed. The d_2 value may be taken from Table 6.5. Formulas for calculating ongoing process capability are given in Table 6.6.

TABLE 6.5

d_2 Values for Ongoing Capability Calculation

N	d_2
2	1.128
3	1.693
4	2.059
5	2.326
6	2.534
7	2.704
8	2.847
9	2.970
10	3.078

N is the number of samples in the subgroups used to calculate the ranges.

TABLE 6.6

Formulas for Calculating Ongoing Process Capability

Statistic	Formula	Description
X-bar	$\sum_{i=1}^{N}(X_i)/N$	Add up all the measurement values and divide by the number of values.
R-bar	$\sum_{i=1}^{N}(R_i)/N$	Add up all the range values and divide by the number of values.
S	R-bar/d_2	Estimate S by dividing R-bar by d_2 from d_2 table.
Cp	(USL – LSL)/(6S)	Subtract lower spec from upper spec and divide remainder by the quantity of 6 multiplied by S.
Cr	1/Cp	Divide 1 by the Cp.
CpU	(USL – X-bar)/(3S)	Subtract X-bar from upper spec, then divide the remainder by the quantity of 3 times S.
CpL	(X-bar – LSL)/(3S)	Subtract lower spec from average, then divide remainder by the quantity of 3 multiplied by S.
Cpk	CpU or CpL	Calculate both CpU and CpL. Cpk is the lesser of the two.
Cpm	$Cp/\sqrt{1+\left(\dfrac{X\text{-bar}-T}{S}\right)^2}$	T = the target or nominal specification value. Use this when the nominal value is not the center of the specification.

The most common errors in process capability calculation have to do with either the normality of the distribution or the way the specification is toleranced. For example, Cpk must be calculated differently for skewed distributions than for normally distributed ones. Also, when the nominal value is not the center of the specification, Cpk is not the best choice for measuring capability. Use Cpm instead. For specification having only a minimum and no maximum, use CpL. Likewise, if there is only a maximum specification, use CpU. Since Cpk is the lesser of CpU and CpL, the Cpk of a unilateral specification will equal CpU or Cpl as applicable.

For processes that are just starting up or for processes that are not normally distributed, use Pp, PpL, PpU, or Ppk. These are not interchangeable with Cp and Cpk. For unilateral specifications (having only a minimum or only a maximum), Pp and Pr are not applicable; use Ppl, PpU, and Ppk as applicable. These are just a few examples of process capability calculations being other than typical. Formulas for calculating all types of process potential are given in Table 6.7.

TABLE 6.7

Formulas for Calculating Process Potential

Statistic	Formula	Description
S_{rms}	$$\sqrt{\sum_{i=1}^{N}(X_i - \text{X-bar})^2 / (N-1)}$$	For each measured value, sum the square of the differences between the value and the average, then divide by $N - 1$ and take the square root.
S_{rts}	$$\sqrt{\sum_{i=1}^{N}(X_i - \text{nominal})^2 / (N-1)}$$	Use either a target value or nominal.
Pp	(USL − LSL)/(6S_{rms})	Same as Cp but use S_{rms} instead of estimating S from R-bar/d_2.
Pr	1/Pp	Divide 1 by the Pp.
PpU	(USL − X-bar)/(3S_{rms})	Subtract X-bar from upper spec, then divide remainder by the quantity of 3 multiplied by S_{rms}. Use S_{rts} to calculate the capability with respect to the nominal value that is not in the center of the specification.
PpL	(X-bar − LSL)/(3S_{rms})	Subtract lower spec from average, then divide remainder by the quantity of 3 multiplied by S_{rms}. Use S_{rts} to calculate the capability with respect to the nominal value that is not in the center of the specification.
Ppk	PpU or PpL	Calculate both PpU and PpL. The lesser of these two is Ppk.

Low capabilities indicate the need to improve the manufacturing process to make it more productive. For most processes, a value less than 1.00 means that you certainly should improve the process. When you have a value between 1.00 and 1.33, capability is usually less urgent, so many companies improve those processes only as resources permit. Above a value of 1.33, process improvement may not have sufficient impact on productivity to be worth improving, although for high-volume products it might.

When Cr or Pr is the capability index used, then lower is better, so higher values indicate the need to improve the process. Typically a Cr or Pr value of 1.0 or higher is when process improvement is absolutely necessary for productivity improvement. At values between 0.75 and 1.0, it is prudent to improve processes when resources permit. When Cr or Pr is less than 0.75, process improvement is usually not cost effective unless the process produces a very high volume of parts. Any time you improve process capability, you are lowering the defect rate. Process capability is what determines the dimensional defect rate. Therefore, it has a major effect on productivity.

There are two ways you can improve process capability: You can either reduce variation or center the process toward the center of the tolerance window. Which one of these you do depends on the difference between the Cp and Cpk (or Pp and Ppk). Whenever the Cp value is not good, especially if it is 1.0 or less, improve the process by reducing variation. If Cp is good and Cpk differs from Cp by more than 20 percent, improve the capability by adjusting the process average. If the Cp is good and the Cpk value differs from the Cp value by 20 percent or less, improve the capability by reducing variation.

In some cases Cp and Cpk are not the best choices to determine the process capability. One such case is when the tolerance limits are bilateral but unequal. An example of this would be a specification like 1.000" + 0.002 – 0.010. In this case, centering the process would mean trying to have the process center be 0.996 which is 0.006 from each end of the tolerance window. However, the nominal value is 1.000" not 0.996", so improving Cpk would actually drive the process away from nominal by 0.004". In this situation, Cpm is a better choice for measuring process capability. It is a process capability index that is considered the target nominal value rather than the process center.

Similar to bilateral specifications, in unilateral specifications you either reduce variation or adjust the process average further away from the specification limit. In the case of unilateral specifications, it depends on how far away from the specification limit the process average is. If the process average is less than three standard deviations from the specification limit, adjust the average away from the specification limit. If it is three or more standard deviations away from the specification limit, reduce variation.

Methods of variation reduction may include but are not limited to the following:

- Improvement of GRR
- Operator training
- Improvement of work technique
- Tool or process modification
- Reducing variation in process parameters or dimensions
- Poke-yoke (mistake proofing)

Methods of centering the average may include but are not limited to

- Adjustment of the process parameters
- Change in setup technique
- Revision of the CNC program
- Centering other related dimensions
- Fixture modification

SPC can be reduced or eliminated if the process is stable and the process capability remains high over a long period of time. Typically, SPC is of no benefit if the CPK remains above 2.33 on a stable process (one that remains in statistical control) for six or more months.

HANDLING, STORAGE, PACKAGING, AND PRESERVATION

Handling can either enhance or reduce productivity. It depends on the routing of the material through the plant, the number of times possession changes hands, and the methods of containment and transportation used. Generally, shorter routing and fewer changes in possession result in more productivity. Containment methods should be chosen to facilitate movement while minimizing the need for handling by personnel. Increased handling can also mean increased opportunity for breakage, loss or misplacement, or even increased product contamination. Any of these hurts productivity.

Storage conditions may deteriorate material or degrade performance if the conditions are not well chosen or controlled. Preservation is simply the means by which this deterioration or degradation can be resisted or even prevented altogether. Packaging is the implementation of preservation. How the product is packaged has much to do with how well it is preserved.

Inadequacies in storage conditions, preservation, and packaging techniques either can cause nonconformances in your material or product or can accelerate the appearance of latent defects. Either way, productivity is affected because such nonconformances or latent defects can reduce the amount of salable product per amount of work effort, which by definition is a reduction in productivity.

TOOLING AND EQUIPMENT

It is not unusual to think, or at least hear it said, that we could be more productive if we had better equipment, more automation, higher quality tooling, and so forth. These are not excuses for a lack of productivity. They are the speaker's assessment of reality. Obviously, not all companies have the latest and greatest tooling and equipment, nor can they all afford to, and so manufacturers do the best they can with what they have. Nevertheless, certain things can improve productivity regardless of the age or condition of your equipment or the quality of your tooling.

Tool availability is a concept that production workers are well familiar with. When a required tool is not available for any reason, time may be wasted looking for it or going to get another one. Hand tools are especially highly mobile and easily carried. Thus, hand tools are more likely to be borrowed by other departments or other operators within a department, or just turn up missing more frequently. Whenever a necessary tool is not available, the employee has to either wait until it is available or compensate for the lack of tool in some way, often with a less-than-adequate substitute. Either of these options can result in nonconforming product, which wastes time, materials, and labor. By doing so, they reduce productivity. This is especially true with hand tools.

One way to deal with this is to attach small chains or strings fastened to the tool at one end and held in place at the other end, while still having enough freedom of movement to allow the tool to be easily used. Department and applicable part numbers could also be stamped or inscribed indelibly into the tool material. Designing tools so that they are easy to see makes them easier to find and harder to lose. This can be done with colored dyes or brightly colored handles on the tools for increased visibility when looking for the tool. Sometimes just having more than one of a particular tool is all that is needed. Anytime a tool is less likely to be lost, less likely to be moved out of place, or otherwise kept more available, it will improve tool availability and therefore help the employees to be more productive.

Companies have implemented various strategies to make sure hand tools are available to the employees who need to use them. Some companies give the employee a toolbox having all the tools the employee needs. This toolbox is assigned to the employee who is responsible for its contents. Other companies have a job box, which is a toolbox assigned to a specific job rather than to an employee. Whoever works on that specific job gets their tools from the job box. A third strategy is to have shadow boards. These are boards that have the outline of each required tool to be hung in its place. The outline is labeled with the tool identification. All the tools are returned to their specific spot and hung on the board at the end of the shift or workday. These solutions to the lack of tool availability all have their merit and have been successfully implemented across various industrial sectors.

Fixtures are one aspect of tooling and equipment that has a major effect on productivity. A well-designed and well-made fixture can make a real difference. Fixtures should be designed to minimize what the operator has to do while still maintaining ease of use. The faster and more easily the operator can install and remove parts from the fixture, the better for productivity. It is all about the number of discrete movements the operator has to make and the ability of the operator to see what he or she is doing.

Certain design features such as thread type, clamping, and angle of view are all factors to be considered. One clamp that attaches the work to the fixture, holding it in several key places, can, if properly designed, hold the part just as well as a group of clamps all holding the work in just one key location each. If one clamp holding it in several places is used, installing and removing the work from the fixture will involve fewer movements by the operator.

Thread type and fixture parts count are also important to productivity. A coarse thread does not need to be turned as many times as a fine thread and therefore is faster to use. A fixture that holds four work pieces may be better than four fixtures that hold one work piece each because it will take fewer operator movements to change all four work pieces.

It is not just the design of a fixture that matters to productivity. The quality of the fixtures matters also. Each fixture will have its own capability, and so the process capability will have to be calculated for each fixture while in use. A fixture associated with a process capability that is below 1.00 should be reworked to improve the capability. Fixtures with process capability between 1.00 and 1.33 should be improved as resources permit.

The amount of variation in the parts caused by the fixture can also cut into the productivity by increasing the defect rate. Fixture to fixture differences must be corrected and minimized or eliminated. Variation caused by a single fixture is also cause for concern if the capability is low.

7

Waste Prevention

Any waste is not productive. In industry there are different kinds of wastes that are contrary to productivity. As the waste increases, the productivity decreases. All kinds of wastes ultimately end up as wasted money; but any waste, regardless of the kind, lowers overall productivity. What is not always so obvious is the variety of different ways that time and money can be wasted. Immediately, material waste comes to mind, but other kinds of wastes are more subtle. One example is *wasted quantity of personnel*, which means an excess in the number of people applied to a specific job. Another example is unnecessary measurements that waste time in labor-hours without really contributing to production.

Production waste and support function waste are two convenient categories on which to discuss wastes in relation to productivity. *Production wastes* mean all manpower, materials, measurement, machining, time, energy, and their associated costs that are directly the result of product realization. That is to say, they are wastes that have to do directly with actual manufacturing or production activities. Support function wastes are wastes from the business functions that support the actual manufacturing or production. These would be wastes in activities like purchasing, engineering, building custodial work, quality activities, and so forth.

PRODUCTION WASTES

Looking at production wastes first, we can examine how these wastes occur and how they decrease productivity. Then we will consider what to do about them. One basic but very important principle to keep in mind is that wasted activity requires means and opportunity, so preventing wastes can be done by eliminating the means and opportunities for the wastes to occur in the first place.

An excess in the quantity of personnel is a waste of labor-hours, but it can also affect the productivity of individuals by allowing them to develop a habit of working more slowly or less efficiently. This degeneration of personal work habits makes them individually less productive. Another undesirable effect is that too many people in one area can lower productivity by physically getting in each other's way. They can also interfere with the social environment of a work area, which affects productivity whether we want to admit it or not. Correctly applied Lean manufacturing can eliminate an excessive quantity of personnel, thereby improving labor-related productivity.

Wasted material includes raw materials, components, parts, subassemblies, assemblies, and finished product. In this book the term *material* includes all of these. Material wastes occur in several ways resulting from different causes and circumstances, including the following most common ones:

- The material is defective.
- The material is incorrect for the job.
- The material was lost.
- The material was damaged.
- The material was installed incorrectly or otherwise incorrectly processed.

All five of these are to some extent controllable or preventable, although the degree of success in controlling or preventing these varies with the circumstances.

Defective material from suppliers is a matter of supplier quality. The usual supplier control activities are intended to prevent defective material from entering your facility in the first place. Effective, well-planned, and well-implemented supplier quality controls can successfully minimize defective material.

Taking shortcuts or going around these procedures in the interest of expediency is not only risky, but it sets a bad precedent. People may think that because it was allowed once, it is acceptable to do it again. Soon it becomes habitual and the supplier quality controls are ignored until a supplier quality issue causes a noticeable productivity problem. Then management wants to know why the supplier quality control wasn't working.

Often the very people who advocated bypassing or deviating from the supplier quality are the ones who want to know why it did not work. This scenario can be somewhat less probable by making it difficult to bypass or deviate from the established procedures, keeping detailed records of who wanted and approved a bypass or deviation, while emphasizing to everyone the temporary nature of the deviation or bypassing.

Internally produced defective material must be dealt with through internal quality efforts. The earlier in the production process the defectives are identified, the less costly they will be in monetary terms, and sometimes their impact on production time will be less. Inspection itself does not prevent defects. It only identifies them. Segregating the defectives keeps them from advancing in the process, but the waste that the defects cause is still present. Generally it costs less to prevent the defects from occurring in the first place. Properly developed design and process failure mode effects analysis (PFMEAs) and control plans are effective defect-prevention tools when applied correctly and conscientiously and kept up to date. Statistical process control (SPC) is another way to prevent defects. Training, ergonomics, quality tooling, and quality materials are more defect preventers. Do not overlook activities like contract review, design review, and qualification testing as ways to prevent defects. These can be effective when done well and at the proper time. Six sigma projects, define, measure, analyze, improve, control (DMAIC), and other quality methods all have their value in increasing productivity by reducing defects and sometimes streamlining processes.

Incorrect material can be an engineering error, a purchasing error, or even a labeling error. Regardless of how the error happens, preventive action is always worth the effort. These are exactly the kind of mistakes that contract reviews, design reviews, design validation, and verification are supposed to prevent. These are all requirements in the ISO 9001 quality management system. Paying only lip service to these requirements or gliding over them during audits provides means and opportunity for waste. It is more productive to have these portions of your company activities be well developed and to use them properly. They are effective ways to prevent incorrect materials from being specified, obtained, and used.

Lost material is a matter of inventory control. It is also a matter of proper labeling, proper storage, and proper distribution. Good practices of logistics and storage, along with positive material identification, are needed to prevent material loss.

Material can be damaged at any point in your activities. Transportation, mishandling, along with loading and unloading accidents, are only one place this occurs. Damage can occur during storage if your building is not in good condition. Leaky roofs, excessively hot or cold warehouses and stock rooms, and improper stacking of pallets are all damage waiting to happen during storage. These are the kinds of wastes that preservation activities are intended to prevent. But to be effective, preservation must be well planned, properly and consistently performed, and not just be given lip service.

Material damage can also occur during production. Poor training, improper tools, mishandling, incorrectly performed operations, and improper testing are all causes of damaged material and therefore lower productivity. Sometimes material can be installed only once, so if it is not done properly the first time, the material is wasted and must be replaced.

Remedies for material wastes require developing and implementing effective corrective and preventive actions to eliminate the waste. Short-term actions can eliminate the means and opportunities that allow the root causes to have their effect, while limiting the amount of impact and preventing the defect from spreading over a larger quantity of product. Long-term actions must eliminate the root causes themselves. Production data analysis, audits, process capability studies, SPC charts, Pareto analysis, and careful observation can all help identify root causes of material waste. They can also verify that whatever corrective and preventive actions you implemented are indeed effective.

Manufacturing process speed is also related to productivity. A press producing parts by a repetitive stamping process has a minimum cycle time, often a maximum number of strokes in a given amount of time. If you exceed this speed, the quality of the parts, the life of the die, or even the condition of the press may suffer. Either of these is a cause of waste. If the press is operated too slowly, you are not making as many parts in any given time period as you otherwise could, and that also hurts productivity. So whether you are too fast or too slow, you are not optimized for maximum productivity. The press and die combination has an optimal speed. This will give you optimal productivity for that press and die combination. This same principle applies to a wide variety of manufacturing processes. When properly performed, all manufacturing processes have an optimal production rate. When the actual production rate is slower than the optimal rate, you are wasting time not manufacturing as many parts as you could be, whereas exceeding the optimal production rate increases the defect rate and so causes more waste.

Wasted time is a big factor in productivity. Wasted time occurs when employees work more slowly than they need to. Idle workers and malpractice also waste time. Time is also wasted by other things, including other wastes. Time is wasted by mistakes, poor work habits, equipment failures, improper sequence of activities, poor planning of other activities, and a host of other causes. Because it is so broad in scope, has so many different causes, and is also the result of other wastes, the reduction of wasted time is necessarily a multifaceted and interdisciplinary effort.

SUPPORT ACTIVITIES WASTES

Wastes from support activities are not directly involved in manufacturing, but they can decrease productivity either by causing wastes or by being wasteful themselves. As already mentioned, skipped or poorly done contract reviews, design reviews, design validation, and verification can cause waste due to incorrect material or by hindering the manufacturing process. If done properly, they also can bring problems to light in the design or prior to assembly of the product so as to enable problem prevention. Such problems can result in poor productivity during production. You can improve productivity by preventing problems with the proper use of these methods and acting upon them prior to production.

Purchasing activities can produce waste, which decreases productivity by a number of ways. Ordering incorrect parts, substandard quality parts, or too few parts are just three ways this can happen. Consider also the impact on productivity when a supplier's delivery is late. Selecting suppliers that will not have issues of quality and delivery is the reason for supplier evaluation. Both the quality department and the purchasing department should take supplier evaluation seriously because of its impact on waste, and therefore, productivity.

Buyers often think they are saving money buying parts or material from the lowest bidder, but it is possible that the time and money spent screening out defective parts and the loss of productivity caused by such wastes of time often costs more than the amount of money saved.

Quality control and assurance activities can have a huge impact on productivity because they have considerable influence on the amount of waste, especially wasted materials and time. Quality rejections trigger dispositions, and the disposition of the defects determines the amount of waste. Even if the materials are returned to the supplier, the time spent receiving, inspecting, dispositioning, and shipping it back out again was all nonproductive time, because it was time and labor spent on activities other than the manufacture of good product.

Quality rejections during manufacture are a waste of material and labor. This is not to say that the quality department produces waste. On the contrary, quality is more about preventing defects than finding them. The sooner quality activities find defects, the fewer defects are made, and the sooner action can be taken to correct and prevent them. Such quality activities actually improve productivity by reducing waste. This is especially true of quality activities that prevent defects from occurring in the first place.

In an effort to reduce overhead, some companies will do only final inspection, with little or no in-process inspection. This actually increases costs because defects are not found until the end of production, when the cost to scrap or rework parts is the greatest and the inspection does nothing to limit the number of parts affected by the root cause. Although final inspection does have value in screening defectives from being sent to the customer, it is more cost-effective to do in-process inspection, when defects can be found sooner. This way, the scrap and rework costs are lower and the problem can be contained to prevent the cause from affecting even more product. In-process inspection is a proven waste-reduction technique.

For productivity, the importance of preventing defects is a good investment; and therefore, having the majority of the quality department costs going toward defect prevention is a wise expenditure of departmental funds. There is real wisdom in tracking quality costs and adjusting the quality effort based on these costs. This will result in greater defect prevention and higher productivity. If the majority of the total quality costs in a company is for dealing with internal and external failures (rejections), then you are not doing enough to prevent defects. If external defects are more than 25 percent of the quality costs, you are not putting enough effort into finding defects before shipping the product.

A majority, typically 50 to 65 percent, of the costs of quality should be defect prevention and detection costs. In no case should the proportion of the quality budget spent on prevention and detection be less than 40 percent of the total departmental budget. Disproportionately lower cost in prevention and detection with higher costs of internal and external failures indicates that the defect prevention activities you are doing are insufficient or ineffective. Furthermore, if more than 30 percent of the quality costs are for internal failures, you need to increase the defect prevention activities to make them more effective. If more than 30 percent of the quality costs are for external failures, you need to put more effort into detection.

8

Productivity and Motivation

Motivation is the driving force by which humans achieve their goals. Motivation is said to be intrinsic or extrinsic. According to various theories, motivation may be rooted in a basic need to minimize physical pain or to maximize pleasure. It may also include specific needs such as eating and resting. Motivation is what moves us to a desired object, goal, state of being, ideal, sense of security, habit, or stress relief. It may even be attributed to less-apparent items such as altruism, selfishness, morality, or avoiding mortality. Employment provides the money needed to obtain basic needs or experience pleasure. An employee's occupational responsibilities can provide specific needs including stability and security, stress relief, achievement of personal goals, or others. The average workplace is about midway between the extremes of high threat and high opportunity. Motivation is a powerful tool in the work environment that can lead to employees working at their most efficient levels of production.

EMPLOYEE MOTIVATION

Workers in any organization need something to keep them working. It may be that the salary of the employees is enough to keep them working for an organization. However, sometimes working just for wages is not enough for employees to stay with a particular company or in a specific job. An employee must be motivated to work not only for the company but at their particular occupation as well. If no motivation is present, employees' quality of work or their work in general will deteriorate. This will inevitably lead to a decrease in productivity. Motivation by threat is a dead-end strategy, and naturally staff is more attracted to the opportunity side of motivation than the threat side.

There are known to be at least three kinds of motivation that affect productivity:

1. The motivation an employee has to stay with the company
2. The motivation an employee has to do their own job
3. The motivation an employee has to be more productive

These can be thought of as three different management goals. Since high rates of employee turnover require more time spent on training and result in a lower amount of productivity, management needs to reduce employee turnover by providing motivation to stay with the company. Employees who are well motivated to do their current job will do it more enthusiastically and probably with more speed. They will also have greater pride in their output and so be more motivated not to make defective product. Such employees are also more receptive to changes that are intended to improve productivity.

All this points to the fact that well-motivated employees are intrinsically more productive. Thus anything that management does to make employees less motivated to remain with the company will hinder productivity. Likewise, if management does not provide motivation for an employee to do their particular job, or in some way takes away motivation to be productive, productivity improvement will be prevented, and productivity might even be reduced.

It has long been recognized that the social situation a worker has in the workplace is very important. Of similar importance are boredom and repetitiveness of tasks. Any one of these—an undesirable situation, boredom, or excessive repetitiveness—can lead to reduced motivation. Job rotation within a work cell is one way to prevent boredom and repetitiveness. If a work cell has five different workstations and five employees, rotating the employees among the workstations is an effective tool for preventing the decrease in productivity caused by boredom and repetitiveness.

Workers can also be motivated when their social needs are acknowledged and when they feel important. Employees who are given freedom to make decisions on the job and are comfortable socially within their informal work groups may exhibit a higher level of motivation. However, in the opinion of some experts, this has been judged as placing undue reliance on social contacts at work situations for motivating employees.

Real-life experience has proven time and again that motivated employees always look for better ways to do a job. Hence, they are trying to be more productive. Motivated employees are also more quality oriented, which is one way to be more productive.

Some believe money is at the lowest level of the motivational hierarchy and meeting other needs is a better motivator. In the opinion of some motivational theorists, praise and recognition are considered stronger motivators than money. Additionally, experience has shown that the motivating effect of money tends to last only for a short period. Nevertheless, most industries still use money as a motivator, and the importance of money as a motivator cannot be ignored. The strength of money as a motivating factor depends mostly on the individual's financial needs and personality. Thus the degree to which an employee will be motivated by money varies considerably from person to person. At higher levels of the hierarchy of motivational needs we find that praise, respect, recognition, empowerment, and a sense of belonging are often more long-lasting motivators than money. Success with these is also variable, again depending on financial need and personality.

Employee motivation is also affected by the way in which jobs are designed and planned. Ideally they should include operator controllability. Operators usually prefer an operation that they have some degree of control over. It gives a sense of ownership, which provides a justification for pride in one's work. If the process has low capability, the job design should concentrate on process parameters. Besides operator controllability, other aspects of a job that increase operator motivation are identifying techniques that separate good performance from substandard performance, identifying cautions to be observed, and mistake proofing, also known as poke-yoke.

Hiring employees who have a high degree of enthusiasm and seem self-motivated is always a good bet. Proper training, having supervisors and coworkers setting a good example, treating employees with respect, and good communication that operates in both directions (to and from management) are other things to consider in employee motivation.

GOAL SETTING

An effective way that productivity can be improved through motivation is to inaugurate a planned program to motivate employees to be more productive that has a clearly defined and well-publicized goal. This goal must be definitive, achievable, and measureable. The productivity improvement goal is determined and a reward for reaching it is given. This program is publicized to the employees, and progress toward meeting the goal is tracked in a way that all can see.

Goal-setting theory is based on the notion that individuals sometimes are more successfully motivated to reach a clearly defined end result. They must also know where the finish line is. Sometimes achieving this goal is a reward in itself. Alternatively, the reward must be attached to the goal. This is especially necessary if the goal originates from outside the person's own needs. To provide adequate motivation, the goal of increased productivity must therefore have some reward attached to it that affects the achievers in a positive way.

The goal is affected by three features: proximity, difficulty, and specificity. An ideal goal should present a situation where the time between the initiation of behavior and the end result is close. A goal should be moderate, not too hard or too easy to complete. While people often want and need a challenge, they also want to feel that there is a substantial probability that they will succeed. The goal must be objectively defined, specific, measurable, and comprehensible to the employees who try to achieve it. A classic example of a poorly specified goal is "getting the highest possible score" or "getting the best possible result." Such goals are neither specific enough nor objectively defined. Such a goal does not provide a definitive difference between where you are and where you want to be. Consequently, neither the amount of effort needed to achieve it nor the location of the finish line can be known. Thus the highest possible score or the best possible result is not as effective a motivating goal as one where the end result, or finish line, is known in advance and the distance from it can be measured.

When a goal is applied to productivity improvement, it must not be a general or abstract concept like "best possible productivity." A productivity goal must be a definite number that can be expressed either as a percent improvement or as an actual productivity measurement level, such as 1,000 pieces per shift. Be sure the goal is actually obtainable and not so far-reaching as to discourage or create doubt about is achievability.

A common strategy for ensuring that a goal is not discouraging or too difficult is to calculate the average and standard deviation of the measurement you are trying to improve. Then create the goal as a new average of one standard deviation more than the previous average (or one standard deviation less if the goal is to decrease).

This statistically calculated goal has proven achievability due to the fact that it is only one standard deviation from the existing average and therefore well within the normal variation. Yet, it is noticeably different enough not to seem too easy to the achievers. When achieved, you can create a new goal that is one standard deviation from that average.

Productivity improvement goal setting, when coupled with the proper rewards, has been shown to improve productivity in a wide variety of industries. Often the reward is time off with pay or pay bonuses, but these are much more effective if done with a high degree of publicized recognition for the employee's effort, and even fun activities on company time. Such ideas usually raise a red flag in the minds of managers who insist such things are either a waste of money or too expensive to do. However, companies that actually do such practices have invariably found that the financial gains of the company due to the increased productivity are greater than the cost of the rewards. Therefore it may be in the best interest of the company to do them.

Be sure to do the following before launching any campaign or motivational program:

- Eliminate most, if not all, management-controllable causes of defects.
- Be sure the employee has substantial control over the process outcome.
- Thoroughly explain the program to maximize employee participation.
- Have management, especially the employees' direct supervisor, be actively and genuinely interested in the outcome. Let the employees witness this interest.
- Supervision must be willing to listen to and try employee ideas without prejudice.
- Provide the staffing and time to adequately perform the studies to verify any breakthroughs, and have a follow-up plan to the program.

Failure, insincerity, or incompleteness in any one of the above will undermine the success of the motivational techniques, and failure to properly motivate can result in failure or reduced success of the productivity goal program.

9

Reliability of the Process and Manufacturing Equipment

Reliability is the ability of something to perform as specified, under the specified conditions, for the specified period of time. When concerned with productivity, it is the reliability of the manufacturing process that is important. Process reliability is simply the definition of reliability applied to a process rather than a product; that is, the ability of the process to produce parts as specified, under the specified conditions, for the specified period of time. The "parts as specified" is just another way of saying parts that meet the design intent and all applicable specifications, or to put it more simply, good parts. The phrase "under the specified conditions," when talking about a process, means all the process parameters and the processing environment. The "specified period of time" can be the time between maintenance intervals or the length of a production run. It may even be the time from one annual factory shutdown to the next one. In other cases it may be essentially continuous. It all depends on your production requirements and company operations.

Reliability is usually measured as a probability that the performance will last for the required time. This is referred to as the probability of success. Alternatively, it is measured as a failure rate (FR), also known as the probability of failure. Higher probability of success, or lower FR, is what keeps the process going.

FR is often used as a means of qualifying a process. When applied to productivity, a process is not ready for production if its FR is too high. The acceptable FR is actually determined by costs, limits of technology, profit margins, design limitations, and volume. Failure rate applies to any material, component, subassembly, or complete device. The failure rate of a subassembly is the total failure rate of its components. Likewise, the failure rate of a complete device or entire manufacturing system is the sum of the failure rates of its parts.

In some industries, higher volume processes are often expected to achieve failure rates of less than 100 ppm, often 64 ppm or even 1 ppm. Processes that are close to the limits of the available technology typically have FR measured in percent.

The reliability of a process encompasses everything included in the process, that is, all the process components. The concept of process components is a broad one. It encompasses all the materials, process steps and material handling, as well as the hand tools, manufacturing equipment, and gauges that are used. It even includes shop floor furniture like benches, chairs or stools, platforms, lighting, room heating, ventilating, air conditioning, and employee training. Like the failure rates of an assembly or a system, the failure rate of a whole process is the sum of the failure rates of all its components. By making a Pareto chart of the various failure rates of the various process components, you can easily see what needs to be improved to bring the most improvement in productivity. To improve the productivity most, you would naturally choose to improve the FR of the process components that have the highest failure rates. However, you must also consider the effect of the components on the product. It may have a greater impact on productivity to choose to improve the FR of the components that have the most direct influence on product quality or a component that appears in the product multiple times, or the most expensive component, rather than one having the highest FR.

Other useful measures of reliability are mean time to failure (MTF) and mean time between failures (MTBF). The longer these time periods are, the less the process is down for maintenance or adjustments. MTF and MTBF can also be used to determine or improve preventive maintenance schedules, or even determine the most cost effective maintenance intervals. MTF is measured as device hours of operation to failure. Remember that devices and operation hours can be transposed. An operating life of 1,000 device hours can be a sample of one piece operating for 1,000 hours or one piece in 1,000 samples failing after only one hour. MTBF is the mean time between failures either of the same failure mode or between different failure modes. The reliability data should state which. Table 9.1 tells how to select a reliability measurement.

Reliability measurement calculations, when applied to processing equipment, are used to determine the preventive maintenance schedule and spare parts lists. Short MTF or MTBF indicates the need for short

TABLE 9.1

Choosing a Reliability Measurement for Processing Equipment

Reliability Measurement	Use When
Product failure rate (FR)	No equipment failure but process produces unacceptable product
Equipment failure rate (FR)	Process jams or stops by itself
Mean time to failure (MTF)	Equipment fails while process is running but not necessarily in continuous operation
Mean time between failures (MTBF)	Equipment failure while process is running continuously

TABLE 9.2

Formulas for Calculating Reliability Measurements

Measurement	Formula	Description
FR	$\displaystyle\sum_{i=1}^{N} FR_i$	Total of all component failure rates
MTF	$\displaystyle\sum_{i=1}^{N} T/N$	T = elapsed time to failure, N = number of process or product runs
MTBF	$\displaystyle\left(\sum_{i=1}^{N} T_i - T_{i-1}\right)/N$	Average elapsed time between one failure and the next
MTTR	$\displaystyle\sum_{i=1}^{N} T/N$	T = elapsed time needed to complete repair, N = number of repairs

preventive maintenance intervals. FR and failure analysis on MTBF data tells what spare parts to keep on hand. When applied to the whole manufacturing system, including people, equipment, handling and storage methods, hand tools, records, and so forth, reliability measurements can help identify weak areas and result in process improvements that can increase productivity. Reliability measurements are calculated as shown in Table 9.2.

The reliability of material handling activities is a matter of getting the right materials to the right place, in the right quantity, at the right time. Failures include such things as wrong material being delivered, delivering the material to the wrong place, delivering too late, or delivering the wrong amount of material. Too little material coming to the input slows or stops the process while waiting for more material. Too much material can get in the way of the process. It is a matter of matching the rate of material input with the rate at which the process consumes material. For best productivity, not only do the two rates have to be equal, but they must be steady. Variation in material flow rate can adversely affect productivity by causing either a bottleneck or idle waiting. Neither is good for productivity.

Calibration and gauge repeatability and reproducibility (GRR) affect the reliability of the measuring processes that must take place as part of the manufacturing process. See the Measuring Systems section in Chapter 3. See Table 3.1 for how to improve measurement reliability by improving the GRR.

Do not forget the effect of employee training on process reliability. Being reliable means performing as specified under the specified conditions for the specified time. As applied to employees, it means the employee must perform their part of the process correctly and consistently in the ambient shop environment for a minimum of eight hours a day, five days a week. Effective and complete employee training will help the employee perform correctly and consistently. Ambient conditions like lighting, noise level, and comfort help the employee to perform consistently and for the required time. Things like repetitive motion injuries, fatigue, eyestrain, and inadequate safety considerations will reduce the process reliability by interfering with quality and quantity of the employee output, and thus reduce employee reliability. This can interfere with productivity.

Fixtures and manufacturing equipment such as production machines are best improved by improving the MTF or MTBF. Hand tools and fixtures typically have adequate reliability, but sometimes there may be room for improvement. If more than one fixture does the same thing but the operator seems to have a favorite, find out why. Operator preference of one fixture over another, even though the fixtures do the same thing, is a clear indication of a reliability issue or a design issue. Operators use the fixture every day, all day long, so they have a real sense of fixture quality, reliability, and ease of use. Talk to the operator about which fixture is favored and why. You may find a way to improve productivity that will pay back in a relatively short time.

Manufacturing equipment like a production machine is more complex. Since you rarely, if ever, can redesign or change such equipment, the key to improving reliability is to prevent unplanned downtime. To improve this you must first measure the amount of unplanned downtime you currently have and know its causes. That is where reliability measurements like MTF and MTBF come in. They can tell you what you need to know.

Determining what preventive maintenance to do by actual reliability tests and statistically determining the preventive maintenance interval from the actual reliability data ensures the most cost-effective and productivity-enhancing preventive maintenance plans. Arbitrary preventive maintenance schedules or preventive maintenance practices based only on perceptions are truly not cost effective. They are not as productive as the preventive maintenance plans based on real data determining preventive maintenance schedules. Using real data requires a sufficient amount of data and some analysis.

To measure either MTF or MTBF, you first must clearly and definitively define what a failure is and how you are going to detect one. Any failure must be either something you can count or something you can measure. Failures that are not countable or measureable are not properly defined. A review of the PFMEA is called for here. You can use it to describe and define failures, or you can add new failures to it, as the case may be. But every failure must be either countable or measureable. Activities such as adding raw material or removing finished parts are not process failures, nor are planned or scheduled adjustments. However, any unexpected or nonroutine adjustments and any incidents where the process produced nonconforming product is considered a process failure.

The MTF of your equipment is the elapsed time from when the equipment is first activated to produce product, to the time it first needs any unplanned adjustment or unplanned maintenance action, or produces an out-of-spec condition or part. Record this elapsed time, and then restart your process. Do this several times to get data for which you can calculate the average. This is your MTF. You do not need to do a special series of process runs for this. You can run the machine as usual, doing whatever you need to do and producing product. Just take your time measurements as soon as the process failure occurs; then fix the process and restart it. Start timing each time you restart the process.

Do not change anything in the process while it is running. If you correct or compensate for any unplanned situation or change the process or process equipment in any way, you have altered the process and nullified the process MTF measurement. Remember this is not about your product. It is about the manufacturing process itself.

MTBF measurement is somewhat different. Instead of measuring the elapsed time from start to first failure as in MTF, you keep the process running and fix or correct each failure as it occurs. Record the elapsed time between one failure and the next. Do this if the failures are the same or not, and regardless of their causes. Then calculate the average elapsed time between the failures. Remember that a *failure* is any unplanned process correction, maintenance, or incidence of nonconforming product being manufactured.

Because longer MTFs and MTBFs result in less downtime, which results in higher productivity, it is in the interest of productivity that you try to improve and therefore lengthen the MTF or MTBF. This can be achieved by preventive action, which may require a modification to the equipment. While redesigning, replacing, or refurbishing the process may greatly enhance the reliability, the cost and time investment might in some cases make this path prohibitive. Exactly what modifications can and should be made is a cost-versus-benefit decision. Although there are situations where this is usually a feasible and wise course of action, admittedly sometimes it is not.

Improving MTF and MTBF are not the only considerations to reduce downtime. Mean time to repair, known as MTTR, is also an important factor. Different failure modes take different amounts of time to fix. While not exactly a reliability measurement, MTTR is influenced by reliability and it definitely affects downtime. One way to reduce MTTR is to reduce the failure rates for the failure modes that take the most time to repair. Another is to have a "crash cart" ready for when a process failure occurs.

On a crash cart, all the tools, frequently needed spare parts, and measurement devices that are most likely to be needed are on one handy movable cart that need only be wheeled to the site of the failure. Preparing a crash cart in advance and keeping it available can save considerable repair time by making it unnecessary for the repair personnel to frequently go get what they need. Everything most likely to be needed is already right there for them. Due to the size of some spare parts, it may be more practical to not include spare parts on the cart, keeping them all in one convenient location near the process instead.

Yet another way to reduce MTTF is to control the environment so it is more process friendly. This means that the ambient environment is to be made less stressful for the components of the process, including all machinery. Generally speaking, cooler and drier is better. Cleaner environments may also help. Vibration reduction or isolation, though more difficult and expensive, may also be warranted.

After MTTF and either MTF or MTBF have been optimized, you can then use them to determine what preventive maintenance to do and how often. Analyzing the failure modes from the MTBF and MTF measurements will tell you what preventive maintenance activities need to be done, whereas the average times themselves will indicate the maintenance intervals. Since the idea of preventive maintenance is to prevent the failures from occurring in the first place, the preventive maintenance intervals should actually be somewhat shorter than the mean elapsed times.

The MTF and MTBF time intervals are averages. Usually they will follow a Weibull frequency distribution. Weibull plotting paper will help you predict the probability of a failure mode happening, and you can judge from that when to do preventive maintenance. The objective here is to prevent the failure mode, so the maintenance activity must be done before the failure mode happens; hence the need for probability determinations.

It is the shape of the curve that is important. A Weibull curve can be anything from exponential to nearly normal to highly peaked. If you do not have Weibull probability plotting paper, you have to approximate the probability curve as best you can or try to determine its shape by making a histogram. Failures due to wear out or random electrical or electronic component failures usually follow exponential probability. Metal fatigue, operator error, many mechanical failure modes, and environmentally influenced failures are often normally distributed or binomial.

In the case of normally or binomially distributed failure times, doing the preventive maintenance before the mean of the time to failure or time between failures will give you a degree of probability that you will prevent the failure mode from occurring in the first place.

The exponential curve is not symmetric about its mean. The mean is actually typically about 36 percent from one end. For exponentially distributed failures, if you do your preventive maintenance for a failure mode about one standard deviation before the MTF or MTBF, it will give a reasonably high probability that you will prevent the failure from occurring in the first place.

The type of failure mode seen during the reliability measurements will tell you what the preventive maintenance action should be. Select the most common failure modes with shortest time to failure or time between failures. These are common failures that need to be prevented. Take such actions as necessary and practical to prevent them. Some failure modes may take a longer time to show. Either their MTF or their MTBF will be significantly longer than other failure modes. Preventing these is accomplished by your longer interval actions, annual, or even something like once every five years.

Do not attempt preventive maintenance on failure modes that occur only once or are obviously unique and isolated occurrences. Instead, have a plan for dealing with these if and when they occur. Also, do not despair that you need to do special reliability resting on your process. Remember that you can run your process for the purpose of production and just record process failures as they happen. Over a sufficient period of time this will yield the reliability data you need and not affect production at all.

Typically, controlling the ambient environment, improving the preventive maintenance, and simplifying the process are effective reliability improvement actions that are more practical in terms of time and money. Sometimes even experimentation and statistical process control (SPC) can result in process improvement.

Another approach to improving productivity by improving the process reliability is to use reliability-centered maintenance, also called RCM. This is a formalized system for determining preventive maintenance from the process reliability data by ensuring that the process does what you need it to do to continuously manufacture product without unplanned downtime or other wasted resources. It was popular in the 1970s and 1980s, but is still as applicable and effective today as it was then. It is defined by the technical standard SAE JA1011, "Evaluation Criteria for RCM Processes." This standard defines the minimum criteria that any process should meet before it can be properly called RCM.

RCM can be used to create a cost-effective maintenance strategy to address dominant causes of equipment failure. It is an approach to defining a routine maintenance program composed of cost-effective tasks that preserve important functions.

RCM is generally used to achieve the establishment of safe minimum levels of maintenance, changes to operating procedures and strategies, and the establishment of capital maintenance regimes and plans. Successful implementation of RCM will lead to optimized maintenance schedules, an increase in cost effectiveness, and improved machine uptime. RCM has been successfully employed to increase process reliability while maximizing the cost effectiveness of preventive maintenance. RCM emphasizes the use of predictive maintenance such as determining MTF or MTBF in addition to traditional preventive measures from the PFMEA.

To begin RCM, answer the following questions completely and clearly. The answers to questions 2, 3, 4, and 6 should already be on the PFMEA. Add more detail in the appropriate columns of the PFMEA if needed to completely answer the questions.

1. What is the process supposed to do (include or create clear documentation of the process performance standard)?
2. In what ways can it fail to provide the required functions?

3. What are the events that cause each failure?
4. What happens when each failure occurs?
5. In what way does each failure matter?
6. What systematic task can be performed proactively to prevent, or to diminish to a satisfactory degree, the consequences of the failure?
7. What must be done if a suitable preventive task cannot be found?

Question 5 deals with the consequences *to the manufacturer* of the various failure modes. Question 7 is often thought of a "plan B" for when preventive action cannot be taken. Do not forget that preventive maintenance is a type of preventive action. Therefore, question 7 answers what can be done if suitable preventive maintenance cannot be done. It may be a matter of replacing part of the process or equipment. In an extreme case, it may be to duplicate the process entirely.

RCM regards maintenance as the means to maintain the functions a user may require of equipment. The initial part of the RCM process is to identify the operating context of the machinery, the goal of the process—a clear definition of the expected results of the manufacturing process. Then review the PFMEA, revising it as necessary.

The important functions of any piece of equipment are preserved by performing routine maintenance. This is successful only when the function, failure mode, and preventive maintenance are all well matched to each other. The function and failure mode must be clearly and completely identified, and the cause of each failure mode must be positively known and confirmed. Levels of criticality must be assigned to the consequences of failure. Some functions are not critical and may be left to run to failure. Other functions must be preserved at all costs.

Maintenance tasks that address the dominant failure causes are then selected. You must be sure the maintenance task directly addresses maintenance-preventable failures. Failures caused by unlikely events or nonpredictable natural occurrences will usually receive little or no action if their risk is trivial or at least tolerable. Nevertheless, when the severity or detectability of such failures is very high in spite of a very low probability of occurrence, RCM encourages (and sometimes mandates) the user to consider changing something that will reduce the severity or detectability to a lower level. The result is a planned maintenance program that focuses scarce economic resources on those items that would cause the most disruption if they are to fail.

10

Implementing Cross-Functional Productivity Improvement

If you work in a company that is more traditional in its culture, has little integration and interaction between departments, or does not operate in an ISO 9001 manner, you may want to begin your cross-functional productivity improvement efforts by choosing a single activity that is easy to implement and produces an easily noticeable productivity improvement. What that activity is will vary from company to company, so you have to determine it for yourself. After that improvement activity is successful, then you can try a larger, more cross-functional project. When that is successfully implemented and people see the results, then an even larger or full-scale cross-functional productivity improvement project can be attempted.

On the other hand, if your company already has a well-integrated cross-functional culture, you may start off with a more complex cross-functional project, but one that is not so large in scope as to frustrate or intimidate anyone. Then, when that is successful, you can try a full-scale, multidisciplinary, cross-functional productivity improvement project. Productivity does not need to be improved all at once. When people see some results, they may be more receptive to the changes necessary for further improvements.

Not all of the activities described in this book will improve productivity by equal amounts. Some will result in significant gains while others will have much more modest results. Besides which activities you are going to do, exactly how things are currently being done in your company is another important factor in determining how much of an impact any particular activity will have.

Regardless of your company's culture, operational system, or management style, any planning of a productivity improvement project necessarily begins with first determining the scope of the productivity you are going to improve. Are you concerned with the productivity of a single workstation? Or an assembly line or department? Or of all manufacturing?

To implement such a cross-functional project, begin with investigating exactly what the situation is in your organization. As previously discussed, investigate in each department what changes need to take place and where. Then allow each department to determine for itself how it can best implement the changes, providing guidance only as necessary. Having the departments develop their own implementation methods provides them with a sense of ownership and responsibility to implement the productivity improvements. This not only helps motivate them to make the changes, but it also helps with overcoming resistance to changes. Since many of the activities are performed in different departments, they can be done simultaneously. If you wish, you can model your cross-functional productivity plan after the one presented in Figure 10.1.

Investigate what needs to be done. Then prioritize which actions will be done in what order. It may not be possible to do them all. Be prepared to compromise somewhat. You will need to prioritize by estimating the amount of impact the actions will have on productivity, but also based on the resources you have. Unrelated activities in unrelated departments may be done simultaneously. Other activities may require more of a cooperative effort, so the management of various departments will need to coordinate their activities.

Be sure to consider auditing and handling/packing/preservation methods in your investigations. Estimate how much productivity improvement you can expect from these. If you expect enough productivity improvement to be worth the effort, try to determine how difficult it will be to implement productivity improvement in them. But do not take action until you have finished investigating.

Next, measure you current actual productivity. To determine this, decide first how you are going to measure it. You may decide on a single productivity measure such as number of units of product per labor-hour. Alternatively you may choose to have more than one input, like the number of units of product per ton of raw material, or per labor head count (instead of hours).

After you have prioritized your productivity improvement activities, decided how you are going to measure productivity, and have measured it, you are ready to implement the productivity improvement project. For a cross-functional productivity improvement project, this should

1. Communicate your productivity improvement intention to all the employees and describe how it benefits them.
2. Develop a cross-functional productivity improvement team comprising all the management personnel whose departments are affected by the cross-functional activities.
3. Train the management team in cross-functional productivity improvement methods as needed.
4. Agree on a method of tracking progress that is understandable by everyone, and post it where everyone can see it.
5. Measure and post your current productivity level.
6. Perform a full-system document review and audit the processes. Make changes as necessary.
7. Make the revisions in the quality system that are necessary for productivity improvement. Show your progress and success to all employees by updating the productivity measurement. Be sure to continue monitoring the productivity at least monthly.
8. Eliminate waste in both production and supporting activities.
9. Implement the usual traditional productivity improvements, and then measure productivity again.
10. Meet with the various supporting departments one at a time to determine what improvement actions to make and how and when to implement them.
11. Have the productivity improvement team plan the sequence of which departments or activities will be improved and which ones can be done simultaneously.
12. Identify and implement additional changes in manufacturing that will improve productivity.
13. To the extent possible using available time, funding, and other resources, measure the mean time to failure (MTF) and mean time between failures (MTBF) of your manufacturing equipment. Then develop and implement preventive maintenance plans to prevent downtime due to equipment failure.
14. Remeasure productivity after implementing each improvement, and plot the measurements on a public chart.
15. Revise any improvement actions that are not effective.

FIGURE 10.1
Example of a cross-functional productivity improvement plan.

include the more traditional methods for improvement in addition to the interdisciplinary cross-functional approach described in this book. Such a shopwide project involving so many different departments naturally requires the cooperation of all the different departments, as well as a sensible plan for undertaking the productivity improvement. This requires that the management of each department buy into the project and the plan to accomplish it, so a meeting with all of them is a good start. It also requires sufficient time to implement. In some companies the company culture may be that cooperation between departments on something that is perceived to be primarily "manufacturing's problem" may be more challenging to obtain.

To help with this management buy-in, have a meeting of all the managers from all the departments that will be affected. Using your knowledge gained from this book, explain the cross-functional approach and give examples of how the various departments can contribute to the increased productivity. Develop a highly visible way to show the increase in productivity and show the progress of implementation.

Cooperation from the various departments is facilitated if they have a sense of really contributing and can see progress being made. To show them progress and help them see their contribution, post a productivity monitoring station in each department. At this station clearly identify what the department is doing to improve productivity, and keep a graph updated so they can see the progress of the productivity improvement project. This productivity monitoring station can be as simple as posting on a bulletin board.

But how do you know where to start? How do you choose the initial activities? Start in your own department. If you are a quality professional, start by checking your system documentation. Look for issues in the quality system. Revise procedures to break the document loops and resolve conflicting directions. Talk to the document users to see what needs to be revised for accuracy and comprehension. Be sure you are not overdocumenting or underdocumenting. Do not over- or underspecify what to do. It is very important to follow any document changes with training the document users and their supervisors.

If you work in sales, start by making sure the contract review process is cross-functional. If you work in engineering, revisit your design verification and validation procedure and consult Chapter 3 for what would make this yield optimal productivity.

As part of cross-functional productivity improvement, you must include the more well-known and traditional productivity improvement methods. These are predominately of a manufacturing nature and can be your entry in the manufacturing portion of the cross-functional improvement project. Because they are more well known and traditional, you may encounter less resistance to doing them.

Next, investigate the management activities that are not directly part of manufacturing, but have more of a supporting role. They would be departmental activities like purchasing, facilities maintenance, design verification and validation, and training methods. These may yield surprising results, although in some companies these might already be optimized for high productivity and no changes may be necessary. Do not forget human resources activities.

Meet with the facility maintenance department and those responsible for preventive maintenance if it is not done by the maintenance department. Discuss with them any issues of the plant or equipment that are detrimental to productivity. If equipment breakdowns occur or if constant adjustments to your manufacturing equipment are necessary, consider developing the preventive maintenance according to the actual equipment reliability.

When the supporting departments' portion of the improvement project is well under way and you have implemented the more traditional productivity improvement methods, you will undoubtedly see real improvement in productivity. The success of the project so far, as well as the newfound familiarity with improvement methods, may inspire more cooperation and reduce resistance to change. Now it is time to start working with your manufacturing processes.

Collect data on inspection time and preventive maintenance, and do gauge R&R evaluations (repeatability and reproducibility) at the workstations. Proper analysis of these data may shed some light on productivity issues to determine the impact they will make. While collecting the data, take the opportunity to gather operator input. Talk to the operators and plant safety person about the ergonomic aspects of the process. They know what can make the process more productive from an ergonomic standpoint. Map out the movement of material and operators to a devise new layout of the flow of materials. Correct all of the counterproductive policies and procedures that you find.

Do not forget to implement the more traditional methods of productivity improvement that are done primarily in manufacturing. If you are doing SPC, be sure to evaluate how effective it is, and then make any changes necessary to increase its effectiveness. Other manufacturing-related things like following up on corrective and preventive actions, handling and use of tooling and equipment, and even employee training must all be addressed.

Without calling it malpractice and without being accusatory in any way, talk to the operators about any deliberate deviations from the standard operational activities. When doing this, mention that you are looking to revise the activities if there is a need to. This may help them to open up and discuss malpractices if they are occurring. Then you can correct them.

Automation may be expensive and take time to implement. However, sometimes there are minor changes to automation of the process that can be of measurable help in improving productivity. The type, amount, and timetable for automation are management-level decisions that must be made on the basis of cost versus benefit. Different companies desire different time periods for payback. Many want it to be a year or less in times of economic hardship. Three to five years is more common in times of prosperity.

Although always limited by time and available money, knowing where you have the highest probability of success and can make the most improvement at once is also an important factor that must not be ignored.

Look at handling, storage, and transporting of work in process. Check on tool availability and improve it as necessary. Measure the GRR and improve it as needed for every critical part measurement or other measurement of major importance to the process. Review operator activities to be sure they are taking place as planned and that the planned way is in fact optimal for productivity. Operator input will be valuable when making these changes. They are often determined by teamwork, where the team consists of engineers and supervisors as well as operators and representatives from subsequent departments or operations.

During the implementation of the productivity improvement project, it is wise to track and publicize the successful achievement of the various implementation milestones. However, just because changes were successfully implemented does not mean the project itself is successful. It is the effects of those changes that will bring about productivity improvement. Measuring the postimplementation productivity and comparing it to the preimplementation productivity tells you if you are successful. The project is successful only when the average productivity after complete implementation is greater than the average was before implementation. Depending on the nature of the changes made, their full effect may not be noticeable right away. How long it takes for the productivity improvement to really show depends on a variety of things like the manufacturing rate, number of inventory turns per month, and completion of learning curves and practice, not to mention the nature of the changes that were made.

In any case it is important to publicly show the success by actual measurement. If everyone sees how successful the productivity improvement project is, other productivity projects in other departments, assembly lines, or work cells will not be resisted as much.

Choose one or more metrics that everyone can understand and a scale on the graph that makes the improvement obvious to see without undue exaggeration. Label the graphs in a way that everyone can understand, and if it seems advisable, include a short explanation of how this is a measure of productivity improvement. If the data are disappointing, find out what went wrong, but remember a modest improvement is still good. Do not overdramatize the effect of the project nor downplay its shortcomings. Honesty and accuracy build trust, and trust overcomes resistance to change.

If your productivity improvement project was for a specific department or product line, the success of the project may encourage similar projects in other departments or with other product lines. However, this may not happen right away. This is nevertheless a good topic to discuss at ISO 9001 management review meetings or other management-level discussions.

If the cross-functional productivity improvement project was broader in scope, it will have taken longer and it will be very important to publicize the results well throughout the company. Results on every product line or every department will usually not be equal, as productivity normally varies with differences in products or departments. Nonetheless, an overall improvement will be easily seen if the project was successful.

In any case, it is important to save the plan and any other related documentation including results for future reference. Document not only what you accomplished, but how you did it. Such information may be of value in years to come when the company has evolved considerably or if further productivity improvements are desired.

11

Overcoming Resistance to Change

Any actions you implement to improve productivity, including corrective or preventive actions, are essentially changes. Every improvement from a less productive to a more productive way of operating is a change. Resistance to change is a universal human behavior. It is found in all countries and among all ethnic groups, races, cultures, and societies. It exists among all professions and trades, and at all educational levels. Resistance to change does not respect age, gender, or health. It is ubiquitous.

It is also highly variable. Resistance to change varies from person to person, and within a person it varies from day to day, hour to hour, and even minute to minute. It is also context sensitive and hormonally influenced. A large number of variables, happenings, and situations affect the amount of resistance that you will encounter. However, good planning, careful observations, and the application of some basic guidelines can significantly reduce this resistance much of the time and in some cases actually overcome it.

Not all changes will be resisted equally, and a high level of management support will significantly reduce resistance. The resistance to change can range from almost insignificant to severely frustrating. Think of this chapter as a toolbox of change-facilitating tools that you can use if and when needed.

To overcome or at least reduce this resistance, first estimate the amount of resistance you expect to encounter in implementing improvements and from whom you expect to receive the resistance. Then look at the degree of productivity improvement you expect from the improvement activities along with the difficulty and complexity they present. At first, when beginning the cross-functional productivity improvement project, try to select an activity with an easy implementation that will make a noticeable impact on productivity. If you must choose between two or more, go for the easiest one first. You can always do the more difficult one later.

Planning the implementation and overcoming resistance to change may seem like a slow process that consumes a lot of resources. A cross-functional approach to productivity improvement may seem like an extremely large project. There are two things to remember: One is that all productivity gains are permanent, so they will pay off in the future, usually much faster than people think. The other is that the improvements need not be made all at once. Every company is different, so what changes you make, the order in which you make them, and how you make them must be customized for your situation. Therefore, any plans you have made to implement productivity improvement changes must necessarily be somewhat fluid and easily adaptable.

Sometimes the change agent is accused of having a hidden agenda or of being power hungry. Both of these perceptions are at times a cause of resistance to change. To be a good change agent, remember that how you present yourself to others is important and can either reinforce this belief or weaken it. Do not always talk in declarative sentences. Occasionally ask questions, whether or not you know the answers. That way, when the person has to give you the answer to the question, you will not seem to be a know-it-all. Elicit opinions from others without expressing your own opinion. That way people will be less likely to think that you have a hidden agenda.

Some degree of skepticism is a good thing. It prevents planning that is irrational or poorly thought out. It forces us not to be impulsive. Never propose a change on impulse. Before presenting a new idea to an audience, an effective change agent will take the time to examine the idea properly, find its weak points, imagine what the arguments against the change will be, and then develop answers for each objection or argument. Be a little bit skeptical yourself until you have answers for the arguments and evidence for the merit of the change. When presenting a new idea to an audience that may resist it, acknowledge that some degree of skepticism is healthy and important. Then explain to the resisting audience the arguments against the idea that you identified and how you answered them. Then ask for other objections and comments. Answer them as well. Keep in mind that the answers need not be given immediately. Some may require careful thought or research. Don't be overconfident or act like a know-it-all. Listen to your skeptics. Some part of what they say may prompt you to make revisions to your idea that may end up being genuine improvements. In some cases, their arguments against making the change may actually be valid and require revised thinking and planning on your part.

However, it is important to be able to distinguish between logical skepticism and emotional skepticism. Logical skepticism brings out specific issues that are logical consequences of the change. They deal only with real issues to overcome, and not with emotions. Logical and valid skepticism is always objective. It does not deal with preferences or unlikely contingencies. It will not contain such words as *might, may, perhaps, could, should, want, desire,* or *prefer.* The presence of these words is a signal that the person is either expressing opinions rather than facts or simply expressing an emotional reaction. These personal opinions and this emotional skepticism are not valid reasons to discard the idea. However, logical and valid skepticism is a good thing and may send you "back to the drawing board" to revise the idea. This is good. It leads to more improvement.

Change is often seen as risky. Resistance to change occurs when the perceived risk of making the change is seen as being worse than the perceived consequence of not making the change. The best way to deal with this is to make a good, logical, but absolutely factual case of why the change is the lesser risk and has the better consequences. Use actual data if you can. Data that come from the resisting party will be more credible to them. Therefore, have them collect and record the data, and if practical, allow them to assist or at least be present during the data analysis. Although it is alright to acknowledge the emotional aspects of change resistance, if you try to make your case only on ideals and promises of reward, or with unsubstantiated beliefs, you will not likely overcome the resistance to the change. You need concrete evidence that is easy to understand and irrefutable. Be cautious here. A factual, data-driven argument will work best only when dealing with data-driven people like engineers, as well as those who work in the fields of accounting and quality. But data and logic alone will not convince others. You need other concrete results that you can show, perhaps a success report from another department, or some experimental results where the resisting party participated in the experiment.

To some extent, resistance to change is a matter of personality. It is also about interpersonal relationships and life experiences. Therefore, the better you know someone, the easier it is to foresee how much resistance you will encounter and how you can overcome it.

If people are insecure about their situation, they may feel an irresistible need to be in control. Other people have autocratic controlling personalities and need to always be in control in any situation. If you have to deal with a very controlling person, you can actually use that to your advantage by having them be the change agent. Here is how it is done: First, talk to them about what is wrong, the difficulties that are hard for anyone to control. Do not tell them what you think. Just ask them what they think. Ask what mistakes others make that is making things worse. Ask what is difficult to control. Ask what has to happen to make things better. Then, ask their advice. A controlling person will gladly tell you how they would implement their solutions and directives that they want to make. Make them feel like they have all the answers and you want to help them fix things. Try to incorporate some of their ideas into the change (even if you do so only temporarily with the intention of removing them later).

Next, ask them how to handle each of the productivity issues that the two of you have agreed on, subtly adding any that you want to correct if they have not already come up. Ask how to handle the issues in the order of the controlling person's enthusiasm for the change necessary to correct the issue. If there are several changes, you may have to talk about them one at a time on different days. Word your questions to include the answer you want. Do this while guiding them with questions to the solution you want them to accept. Have them see that by implementing the change, they are controlling the issue. This not only satisfies their need to control and eliminates their resistance, but it also helps you to implement the change by using an experienced controlling person to implement the changes.

Another personality situation that causes resistance to change arises when a person does not like the person or people associated with the change. If a person has a low opinion of the judgment of the one proposing the change, resistance may arise simply because of the one who is proposing it. If they do not like the group or individual implementing the change, their dislike will be perceived by them as distrust, and they may rationalize excuses for not accepting the change. In this situation, if the person or group proposing the change is the reason for the change resistance, then just have someone else present the change. The proper parties can get their due credit later. Have the acceptable proposer participate in the implementation in some way.

Sometimes people seem to be more close-minded than others and frequently resist change. Change is more difficult for them if they have not personally experienced something. Consequently they have difficulty seeing the merit in another way of doing something. To them, change may be a terrible journey into the unimaginable, which to them is a complete unknown, a void, or a boundless abyss, and therefore, confusing, frightening, and most difficult to deal with.

To overcome this kind of resistance, you must eliminate the need for them to imagine something that they have not experienced. Always demonstrate, rather than explain. Build a working model. Show them pilot programs and realistic examples. Do anything you need to do to avoid them having to imagine a situation. In most cases their resistance to change will then decrease at least somewhat.

Along with the personality issues, other variables of change resistance are common, at least some of which will be found applicable to any one person who is resisting change. Understanding these variables and how to deal with them is critical to successfully implementing any change.

We are by nature a social species. We seek familiar people and want to be around those with whom we have steady relationships. Implementing a change can upset the social structure of a department, an assembly line, or a work cell if it moves people around or away from the area entirely. We naturally resist upsetting our social structure. The only things you can do then are to verbally and by your actions honor the work and successes of the group while also emphasizing the contributions of the individual or individuals who will have to go somewhere else. Have everyone in the group participate in this recognition. This way, leaving them will seem less like a betrayal and so become just a little bit easier.

Upsetting the social structure can also give a person anxiety about what their new status will be. Providing a limited amount of emotional support and helping a person fit into the new group dynamics may reduce this and so reduce their resistance to the change.

Another cause of change resistance is a feeling of incompetence or low self-esteem. People are afraid that they cannot handle the new way. It may be true in some cases, but often there is no basis to believe it. In a majority of situations, incompetence is not a real issue. While training programs and emphasizing teamwork are good ways to handle this, do not stop there. Create a test or trial *of the change* in which the resisting party helps you "debug" the change. Let them tell you the best way to do the change. This will shift their self-perception from that of incompetence to that of competence by making them feel they have some expertise and input in developing the change. This may help raise their self-esteem and their confidence in handling the change.

Timing of a change is also important. Holidays and periods of peak vacations times are not when changes should be introduced because the state of mind that people are in is more focused on tradition and personal agenda. Immediately following a newsworthy natural disaster or before a major storm is also not a good idea because people are already in a highly stressed state of mind, and introducing change just adds more stress. Hence their resistance to change will be greater.

In addition to the aforementioned advice for dealing with the various causes of resistance, a few more strategies will help overcome resistance. One is to use a team approach. It may be wise to have the resisting person be a part of the team. Teams share expertise and spread responsibility, so each team member feels less self-conscious. They also supply peer pressure to reach agreement, and that can help reduce or eliminate resistance to change. Another purpose of the team is to oversee the implementation of the productivity improvement project.

Establishing the team must be done carefully. Choose members that represent all the stakeholders who will be involved in the change. Select from both management and labor if both are affected by the change. Pick the team members on the basis of their people and leadership skills. They and their supervisors must consider the team membership and activities as part of their regular jobs. For the team members, establish some metric that will definitively and accurately portray the project implementation as it relates to their particular jobs. Post the progress publicly for all to see.

Another strategy is to clearly define the overall goal and implementation plan. Keep everyone focused on the goal and on track with implementation. Having some flexibility in the implementation will enable you to adapt to unexpected roadblocks. You may have to allow and implement less-than-ideal changes or even ineffective ones. Do so with the implementation team's understanding that any action taken to improve productivity is subject to review and evaluation afterward. Point out to the team that any change that is not yielding the best results can be tweaked, replaced with a better action, or even canceled. This not only makes tweaking and correcting action part of the plan, but it also allows poorly crafted or ineffective actions to show themselves. By allowing those actions in the first place, you have also allowed those actions to show their inadequacy and created a learning experience that is inadequate, which can help overcome the resistance to changes that you really wanted to make.

A third implementation strategy is to consider a wide variety of methods when planning how to do the implementation. This broad-based thinking can ensure that the best methods have been considered. Different methods take different amounts of time. Do not make the mistake of selecting a specific change or a method of implementation just because it is shorter or faster. Allow adequate time to do the best method. Do not underestimate the amount of time that implementation will actually take.

Finally, have a plan to ensure compliance to the changes. Constant vigilance, enforcement, and follow-up are needed. Without ensuring compliance daily, the resistance to change may gradually win out, and the change will be neither permanent nor effective. Supervisors, auditors, inspectors, trainers, and the change team have an important role here to be sure that the change eventually becomes the normal and habitual way of operating.

Identify milestones in the project implementation and celebrate each milestone with the whole group that is affected by the change. Make these celebrations occur at important milestones, and do not trivialize them. Remember that cross-functional productivity improvements are long term, usually permanent, lasting month after month and year after year. So a little and money spent celebrating and recognizing achievement is well spent. It is actually an investment in future improvements.

Glossary

Analysis of Variance (ANOVA): A statistical technique for identifying and comparing the causes of variation.

AQL (Average Quality Level): The average proportion of defectives in a continuing series of lots manufactured by the same producer using the same manufacturing. AQL sampling plans are designed to minimize the rejection of acceptable samples and so are often used by producers when inspecting their own product. In many sampling plans, AQL determines the sample size, with smaller AQLs requiring more samples.

Assignable Causes: Causes of variation other than the normal random variation found in nature. Assignable causes are attributable to specific causes.

Attribute Gauges: Gauge capable of determining only if a characteristic is acceptable or rejectable with no measured value being taken or recorded.

Attribute Inspection: Inspection done on a pass/fail basis without an actual value being recorded. Attribute inspection checks the presence or absence of a characteristic or checks whether or not a characteristic is within tolerance.

Audit: A review of the procedures, their implementation, and records to verify if the procedure is being correctly carried out, to assess its merit, and to identify opportunities for improvement. Audits may be internal, second-party, or third-party. Second-party audits are performed by the company's customer.

Bias: The amount by which the average of a particular measurement calculated from values measured by the same device differs from the actual value of the measurement; essentially the same thing as calibration error.

Binomial Distribution: A frequency distribution similar in shape to a normal curve but applicable to attribute data that is counted rather than measured.

Control Chart: A graph that shows whether or not a process is in a state of control by having its measurements remain inside statistically determined limits and by not showing any nonrandom pattern.

When a control chart does not show that the process is in control, it is a signal that the process needs to be examined to determine the assignable cause and a process adjustment may be necessary.

Control Limits: Values on a control chart that, when exceeded, indicate the need to investigate a change in the process and possible need for process adjustment.

Control Plan: A document consisting of a plan that prescribes which process and product characteristics need to be kept in control and how they are going to be controlled. The control plan prescribes what is measured on which characteristics, the measurement methods, and how to react to those measurements when they are out of tolerance, so as to keep a process in control and to prevent the process from producing nonconforming parts or products.

Counterproductive: Causing a worsening of productivity, similar to non-productive in that there is no direct effect on productivity.

Cp: A measure of process capability calculated as the difference between the specified maximum and the specified minimum, which is then divided by 6 times the standard deviation. For Cp the standard deviation of ongoing processes is usually calculated as the average of the ranges divided by d_2 (R-bar/d_2 method).

Cpk: A measure of ongoing process capability for a particular characteristic equal to the difference between the average value of the measured characteristic and the closest specification limit, which is then divided by 3 times the standard deviation. In Cpk the standard deviation is usually calculated by the R-bar/d_2 method. From Cpk, both the process characteristic acceptance rate and the defect rate can be predicted. Cpk is the lesser of CpL and CpU.

CpL: The capability of a process in relation to the lower specification and having the standard deviation estimated by R-bar/d_2.

Cpm: A measure of process capability measurement that is referenced to the nominal value regardless of whether or not the nominal value is at the center of the tolerance.

CpU: The capability of a process in relation to the upper specification and having the standard deviation estimated by R-bar/d_2.

Cr: A measure of process capability that tells you what percent of the total tolerance window is being used by the process. Higher values indicate excess variation and high effect rates. Cr is calculated as 1/Cp.

Cross-Functional: Applying to or making use of expertise from different functions within an organization.

Designing of Experiments (DOE): A statistical way to carry out experiments so as to maximize the knowledge gained and minimize the number of experimental runs. This is used as a means of process improvement by analyzing the experimental results. At a basic level the DOE tells you the level of significance that each variable tested has on the process, how the variables interact with each other, and the optimal values at which to set each variable to produce optimal results.

Design Validation: The manufacture of the newly designed product sample to ascertain that the design can be built and will function as specified.

Design Verification: A review of a new design to ascertain that it meets all of the applicable requirements and regulations.

Distribution Curve: A mathematically constructed graph that matches the frequency distribution of measured values of a particular characteristic. Distribution curves are approximated by histograms.

DMAIC: An acronym for define, measure, analyze, improve, and control. It is a quality improvement methodology used often in aerospace but applicable to any manufacturing. The quality improvement and process optimization it provides can increase productivity.

Downtime: The amount of time a process is not actively producing product, regardless of the reason.

Ergonomics: The study of the interaction between a person and their immediate environment and how they affect each other. Among other things this includes the person's total physical situation including equipment, location, ambient environment, and the task being performed.

Exponential Distribution: A distribution curve having a highly skewed, nonlinear, and characteristically shaped frequency graph where a majority of the measured values are at one end of the curve, with a very low frequency of values at the other end. It is applicable to random failures of electronic components and the wear of a single mechanical tool or part.

External Audit: An audit in a company that is performed by auditors who are not employees of the company being audited. All external audits are either second- or third-party audits.

Failure Rate (FR): The proportion of failures from the quantity present in a process or product; also the probability that a process will produce a defective part.

Gauge Bias: The degree to which a gauge is out of calibration.

Gauge Linearity: The degree to which the calibration of a gauge remains constant over the entire measurement range of the gauge.

Gauge Repeatability and Reproducibility (GR&R or GRR): The combined effect of the repeatability and reproducibility of a gauge.

Gauge Stability: The ability of a gauge to stay in calibration over time. It is calculated as the absolute value of the maximum bias that occurs over time.

Histogram: A frequency graph of all the measured values of a set of measurement data when sorted from lowest measured value to the highest. As the number of data points increases, the histogram approaches the distribution curve of the data.

Interdisciplinary: Applying to, involving, or otherwise affecting various different departments or fields of knowledge that must interact to produce the desired result.

Internal Audit: An audit of an activity taking place in a company that is performed by the company's own employees. All internal audits are first-party audits.

ISO 9001: An internationally recognized organizational management system featuring, among other things, continuous improvement, meeting customer requirements, and easily verifiable quality status.

ISO 10011: An internationally recognized standard for calibration systems.

IX-MR Chart: A type of SPC control chart having a sample size of one, in which the individual measurements of a characteristic are plotted on a graph. Also plotted is the moving range (MR).

Labor Efficiency (Labor Productivity): Productivity calculated as quantity of acceptable output compared with labor-hours input, e.g., 100 pieces/hr.

LCLr: Lower control limit on a range or moving range SPC control chart.

LCLs: Lower control limit on a standard deviation SPC control chart.

LCLx: Lower control limit on an X-bar or IX SPC control chart.

Lean Manufacturing: A planned program of maximizing labor efficiency and maximizing productivity.

LTPD (Lot Tolerance Percent Defective): Another term for RQL.

Maintainability: A measure of labor-hours necessary to maintain a process exhibiting a given failure mode.

Material Productivity: Quantity of acceptable output compared with amount of raw materials or components input, e.g., 100 pieces/pound of steel.

Mean Time between Failures (MTBF): The average elapsed time from one failure condition to the next during a process or between the failure of one device and the failure of the next device in a reliability test.

Mean Time to Failure (MTF): The average elapsed time from the activation of a process or product function to the first sign of failure, regardless of failure mode.

Mean Time to Repair (MTTR): The average amount of labor hours it takes to repair any given process failure mode to the extent that the process can continue.

Measurement System: All the items and aspects of measuring a characteristic that are in any way directly involved with the measurement process, typically including the parts, gauge, fixtures, personnel, work cell design, data recording method, and operator technique.

MR (Moving Range): The difference between a measured value of a characteristic and the measured value of the same characteristic on the next consecutive part.

Multidisciplinary: Applying, involving, or otherwise affecting more than one department or field of knowledge with or without interaction being necessary.

Nonproductive: Not contributing to productivity.

Nonrandom Variation: Normal random variation that has no particular cause or causes and does not exhibit any pattern.

Normal Distribution Curve: A graph depicting the normal bell shape of the distribution of measurements. It is applicable to many mechanical failures and characteristics and almost adimensional characteristics when no assignable cause is present.

Np Chart: A type of attribute SPC chart that plots the number of defective parts in the sample. Each defective part is counted only once, even if it has more than one defect on it.

Pareto: A type of bar graph showing the frequencies of occurrence or relative proportions of occurrences of defects that has been sorted from most frequent to least frequent.

P Chart: A type of attribute SPC chart that plots the percent of defective parts in the sample regardless of the number of defects that occur on an individual part.

PFMEA: An acronym for process failure mode effects analysis. It is a list-ing of all process steps, the different ways the steps can fail, the probability and severity and detectability of the failure, the risk priority value, and any actions taken to prevent the failure from occurring, thereby increasing productivity. From the PFMEA, the control plan is generated.

Pp: A measure of process potential similar to Cp but applicable to a pro-cess at start-up rather than when it is ongoing. It indicates the potential of a process to produce good parts when all sources of variation, including assignable causes, are at work in the process. It is calculated the same as Cp except the standard deviation is calculated by the route mean square method.

PPAP (Preproduction Parts [or Process] Approval Process): A process by which new products or new processes are proven to operate with a sufficient level of quality. Implementation of corrective and preventive action or other problem solutions identified during the PPAP result in productivity improvement.

Ppk: A measure of process potential similar to Cpk, but applicable at start-up when all sources of variation including assignable causes are at work affecting the process. It is calculated the same as Cpk except that the standard deviation is calculated by the route mean square method. It is the lesser of PpL and PpU.

PpL: Similar to CpL except it is applicable to processes at start-up and the standard deviation is calculated by the root mean square method.

Ppm: Parts per million; often used as a measure of failure rate or pro-portion defective.

PpU: Similar to CpU except that it is applicable to processes at start-up and the standard deviation is calculated by the root mean square method.

Pr: A measure of process potential that tells you how much of the tol-erance is being used by the process at start-up when all sources of variation including assignable causes are at work affecting the process. It is calculated as 1/Pp.

Precontrol Chart: A type of control chart that warns the user when a process may be about to go out of control, so that the user can adjust the process before it goes out of control.

Preservation: Packaging to reduce or prevent deterioration, degradation, or expiration of the product.

Probability Curve: A graph depicting the probability of occurrence for a measured value or count. Mathematically it is calculated as a definite integral of the frequency distribution curve.

Process Average: The average measurement of a characteristic calculated from a sufficient number of subgroup samples to give a valid estimation of the average value that can be expected when the process is running.

Process Capability: A measure of the ability of an ongoing process to continuously manufacture the characteristic being measured within the specified tolerance. It is calculated from an estimate of the standard deviation as a measure of process variation that does not include any special causes of variation. It is measured by Cp, Cpk, Cpm, etc.

Process Centering: Adjusting a process so that the process average is midway between the upper and lower specification.

Process Parameters: All the measurable and controllable characteristics of a process.

Process Potential: A measure of the potential ability of a process that is just starting up or not yet optimized. It is measured prior to removal of assignable causes of variation. Its variation is measured by actual root mean square calculation of the standard deviation. Process potential is expressed as Pp, Ppk, etc.

Productivity: The quantity of acceptable output compared with the chosen input or time period.

R Chart: A chart of the ranges of values within a single sample subgroup.

Reliability: The probability that a process or product will function as specified, for the specified period of time, under the specified conditions. The reliability of a manufacturing process has a positive correlation to its productivity, but is only one of several factors that affect productivity.

Repeatability: The amount of measurement variation seen when a series of measurements of a single characteristic are made by the same operator using the same gauge and technique. Also called EV or Equipment Variation.

Reproducibility: The variation in the average values of a group of measurements made with the same gauge on the same characteristic by different operators. Also called AV or Appraiser Variation.

RPN (Risk Priority Number): A number that indicates the amount of risk in having a particular failure. It is calculated from the probability, severity, and detectability rating on a PFMEA.

RQL (Rejectable Quality Level): The average proportion of defectives in a continuing series of lots that would be detected by a given sampling plan designed to minimize the risk of accepting samples that should be rejected. RQL-type plans are best used at incoming inspection.

Saleable: Ready and fit to be sold to the customer.

Six Sigma: A planned methodology for improving products and processes. Such improvements can increase productivity.

Special Causes: See Assignable Causes.

S_{rms}: Standard deviation from the mean when calculated by root mean square method rather than being estimated by R-bar/d_2.

S_{rts}: Standard deviation calculated by deviation from a target value rather than the mean.

Statistical Process Control (SPC): A statistical technique for keeping a controlled process in control by identifying when the process needs to be adjusted.

Subgroup: A sample consisting of a predetermined number of consecutively manufactured parts that is treated as a single sample for control-charting purposes.

Throughput: The amount of material passing through a process or a process step.

Tribal Knowledge: Knowledge, often undocumented, known to all the people who routinely perform the same task.

U **Chart:** A type of SPC chart that plots the total number of defects found in a sample, rather than defective units. If more than one defect is found on a single unit, each defect is counted separately.

UCLx: Upper control limit on an X-bar or IX SPC chart.

UCLr: Upper control limit on a range or moving range SPC chart.

UCLs: Upper control limit on an S-type SPC chart.

Variables Gauge: S measurement device capable of taking an actual measured value of a characteristic that can be recorded.

Variables Inspection: Inspection performed by measuring an actual value and comparing the value with the specification and tolerance.

Weibull Distribution: A kind of frequency distribution often used in reliability and calculated by a scale parameter α, a shape parameter known as β, and a location parameter known as γ. These parameters, calculated separately, define the Weibull distribution, which describes the frequency distribution of a variety of failure modes. Weibull probability may be obtained from plotting the frequency distribution on Weibull probability paper or by integrating the frequency distribution.

Weibull Probability: The probability of failure mode occurrence calculated from Weibull probability plotting paper or by the Weibull distribution.

Work in Process (WIP): All material currently in the production process that is destined to become product.

X-Bar and R Chart: A type of control chart plotting the average value of a sample subgroup and its range.

Recommended Readings

Amsden, Robert, Davida Amsden, and Howard Butler. *SPC Simplified*. Quality Resources, New York, 1998.

Automotive Industry Action Group. *Measurement System Analysis*. Southfield, MI, 1990.

Blank, Ronald. *The Basics of Quality Auditing*. Quality Resources, New York, 1999.

Blank, Ronald. *The Basics of Reliability*. Quality Resources, New York, 2004.

Enrick, Norbert Lloyd. *Quality Control and Reliability*. Industrial Press, Inc., New York, 1977.

Harrington, H. James. *Business Process Improvement. The Breakthrough Strategy for Total Quality, Productivity, and Competitiveness*. McGraw-Hill Book Company, New York, 1991.

Head, Christopher W., and Carl G. Thor. *Handbook for Productivity Measurement and Improvement*. Productivity Press, New York, 1993.

Juran, Joseph M., and Frank M. Gryna, Jr. *Quality Planning and Analysis*. McGraw-Hill Book Company, New York, 1980.

Juran, Joseph M., and Joseph A. DeFeo. *Quality Control Handbook,* 6th edition. McGraw-Hill Professional, New York, 2010.

McDermott, Robert E., Michael R. Beaureguard, and Raymond J. Mikulak. *The Basics of FMEA*. Quality Resources, New York, 1996.

Torok, Robert M., and Patrick J. Cordon. *Operational Profitability: Systematic Approaches for Continuous Improvement,* 2nd edition. Wiley, Hoboken, NJ, 2002.

Index

Page references in **bold** refer to tables.